GERMAN ANTI-TANK AIRCRAFT

A Ju 87 G-1 of Panzerjägerkommando Weiss in Russia with the depiction of a Soviet T-34 tank on its engine cowling.

TANK HUNTERS & ASSAULT AIRCRAFT OF THE LUFTWAFFE

MANFRED GRIEHL & JOACHIM DRESSEL

Schiffer Military/Aviation History
Atglen, PA

The Junkers J 8 from 1918 was the prototype for the J 10 (CL 1) fighter-bomber. Forty-three of these two-seat low-wing monoplane aircraft were built with the proven corrugated light metal skin. The Daimler-D-IIIa engine, with its 160 horsepower, gave the aircraft a top speed of 190 km/h with takeoff weight of 1,135kg.

SOURCES/PHOTOS

- **G.Aders/W. Held, "Stukas, Jagdbomber, Schlachtflieger"**
- **C. Bekker, "Angriffshöhe 400ST1"**
- **G. Brütting, "Das waren die deutschen Stuka-Asse 1939-1945"**
- **F. von Bülow, "Geschichte der Luftwaffe"**
- **R. Cescotti, "Kampfflugzeuge und Aufklärer"**
- **E. Creek, "Jet Planes of the Third Reich"**
- **O. Hahn, "Deutsche Geheimwaffen 1939-1945"**
- **D. Irving, "Die Tragödie der deutschen Luftwaffe"**
- **B. Lange, "Typenhandbuch der deutschen Luftfahrttechnik"**
- **M. Pegg, "Luftwaffe Ground Attack Units 1939-1945"**
- **F. von Poturzyn, "Luftmacht"**

Translated from the German by James C. Cable.

This book originally appeared under the title,
Flugzeuge gegen Panzer,
by Podzun-Pallas Verlag, Friedberg.

Printed in the United States of America.
ISBN: 0-88740-520-7

We are interested in hearing from authors with book ideas on related topics.

PRUSSIAN REAR GUARD
As the sun breaks through a spring storm front over the Prussian frontier, a pair of Henschel Hs 129 B-3/WA strafers from 14.(Pz)SG 9 sweep down upon an advanced column of Russian T-34 tanks. The wingman has just hit his target with a round from his 75mm PaK 40L. A second firing pass with his two 20mm guns will finish off any surviving Soviet crewman. Developed to stop the latest models of heavy Russian tanks met in late 1944, the hefty PaK 40 was the ultimate weapon carried by Henschel's resilient "airborne can opener." Even though fewer than thirty Hs 129 B-3 types were fitted with the recoiless gun which carried just twelve rounds, the weapon system quickly proved to be the most deadly panzer-killer on the eastern front. One 26-pound armor-piercing round was enough to knock out any armored vehicle or fortified ground target. However, the B3/WA was severely restricted in maneuverability and range, and there were far more targets that PaK 40 armed Henschels. Combat attrition overcame the Henschels of SG 9 well before the war ended.

Archival Credits
B. Lange archives
Nowarra archives
Archives of the Forschungsgruppe
Luftfahrt-geschichte d.V.-Mainz
Bundesarchiv Freiburg
Deutsches Museum, Munich
Heinkel archives
IWM – London
Luftarchiv
Messerschmitt-Bölkow-Blohm
NASM – Washington D.C.
Filley collection
Kössler collection
Petrick collection
Radinger collection
Spork collection
Selinger collection
Authors' archives

Published by Schiffer Publishing Ltd.
77 Lower Valley Road
Atglen, PA 19310
Please write for a free catalog.
This book may be purchased from the publisher.
Please include $2.95 postage.
Try your bookstore first.

Introduction

The development of fighter-bombers, anti-armor and close air support aircraft increased dramatically during the second world war. Starting with aircraft which were hardly armored at all, the trend continued toward high-performance and much improved aircraft with large-caliber weapons and accurate droppable munitions which were well suited to each particular mission.

Dive bombers of the Hs 123 and Ju 87 variety began to be replaced beginning in 1943 at the latest by fast and more maneuverable Fw 190's. This aircraft was also used to try to kill tanks from the air, but the future belonged to specialized aircraft. This is why heavily armored ground attack aircraft, such as the Hs 129, were conceived to engage even the heaviest of tanks on the battlefield.

In the end there were jet aircraft as well as, in response to the urgency of the moment, numerous hastily-configured makeshift fighters such as the Bü 181 or the Go 145.

But even the high-performance aircraft such as the Me 262 Blitzbomber could only change the situation to a moderate degree, but they did point the way for the varied post-war construction types.

Below: The Junkers A 48 was developed from the K 47 and primarily served research purposes. The Ju 87 came to be based on the results of this research. Toward the end of the war, the remaining models were taken over by the Luftwaffe as training aircraft.

Below:

The first Heinkel He 50 A's went to the DVS in 1933 for dive-bomb training. When the first dive-bombing combat units were formed in 1935, their inventory consisted primarily of this aircraft. After the Hs 123 and Ju 87's were introduced, these were assigned to flight training.

Early Developments

From the Ju A 48 to the He 50

The Junkers A 48 two-seat aircraft was begun in 1927 under the leadership of the Plauth/Pohlmann team. This was based on a construction model developed by Pech. Of the ten aircraft built, more than four crashed, taking the life of the leader of the Junkers-Motorenbau in one case.

The D-2185 served as a test bed for dive brakes and initially had a simple tail unit. This prototype, which was driven by a Siemens "Jupiter IV" engine, had a maximum speed of 265 km/h. The results of this testing later influenced the development of the Ju 87 and gave Junkers quite an advantage in this area. The A 48 was later utilized in the Luftwaffe as a training aircraft rather than a ground attack or anti-armor aircraft.

In 1932, the Heinkel firm developed a robust biplane, the first land model of which came to be the He 50 aL two-seater capable of steep-angle dives. The first of three SAM 322B radial engine aircraft were sent to the Deutsche Verkehrsfliegerschule in 1935 as He 50 A's. In 1936, those aircraft belonged to the first equipment of the secretly created I. Gruppe of what was to become Stukageschwader 162 in Schwerin. The cover title of "Fliegergruppe Schwerin" was dropped on September 5, 1937, at which time the unit had 24 He 50's. Because the He 50 was somewhat awkward, dive-bombing training began to be conducted with He 51's, which could be fitted with five small explosive bombs, and which was later called into action in Spain against ground targets and even against light tanks delivered by the Soviet Union. The He 50 was next given to training squadrons and deployed to the Eastern Front within the framework of the creation of night fighter units. On December 1, 1943 the 1. and 2./ NSG 11 still had a total of 24 of these aircraft in its inventory, of which 19 were airworthy. The last deployments of the He 50 were in the Fall of 1944 when Luftflotte 6 flew sorties against motorized rifle regiments.

Opening ceremonies of the Neubiberg airfield. Parked He 50's can be seen in the right of the photo.

Because of the He 50's clumsiness, dive-bombing schools began to use the Heinkel He 51 as their trainer. For these purposes, it could be fitted with up to five training bombs. In Spain, 135 of these aircraft saw action with the Legion Condor, at first as fighters. Once the Soviet Polikarpov I-15 and I-16's had been introduced to the theatre, the He 51's had to be withdrawn from the fighter units because of their low speed, and were then used as low-altitude attack aircraft with considerable success.

Beginning in 1943, the He 50's still on-hand were given a new mission as nighttime fighter-bombers on the Eastern Front.

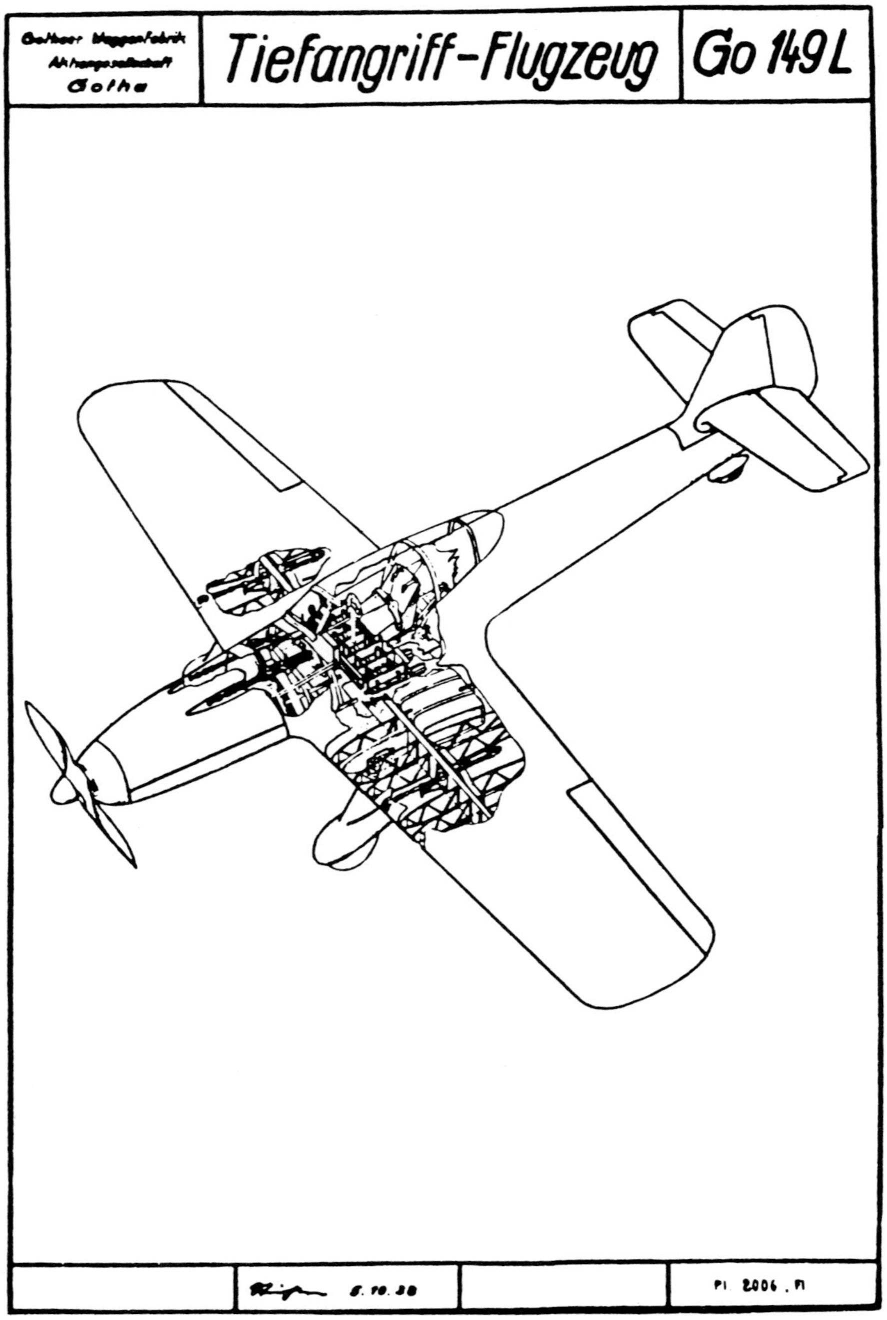

Above: The Go 149 was offered to the Luftwaffe as a single-seat trainer and had a 240 horsepower Argus As 10C engine. The Go 149 V1, registration D-EGWF, made its maiden flight in 1936 and took part in the Frankfurt air races in 1938 and 1939.

After the Go 149 had been declined as a trainer aircraft, the first plans for the Go 149L single-seat fighter were drawn up for homeland defense and night-fighter duties, but this version was never built.

Hs 123 and the Bv 137

In 1935, the Hs 123 represented the first single-seat dive and ground attack aircraft created especially for the Luftwaffe. The final two test versions were delivered to troop units in May and June of 1938.

Because the Luftwaffe leadership believed that after 1938, these aircraft would play a large role in any conflict as low-level attack aircraft, the Hs 123 was stricken as a front line aircraft and therefore the entire construction setup, templates and models were hurriedly scrapped. Only 115 aircraft left the HFW (Henschel Flugzeug Werke) production facilities; another 129 single-seaters were to follow from the AGO facilities, produced there under license. The constructive improvements to the Hs 123 as proposed by the Henschel works found no support from the RLM.

The modified version was to have:

a.) BMW 132 K engine with a VDM propeller

b.) enlarged fuselage tank

c.) enclosed cockpit

d.) two additional MG 17 machine guns with 500 rounds each, and

e.) an increased bomb capacity to eight 50-kilogram loads or four SC 50's and a maximum of one 500 kilogram high-explosive bomb.

However, the decision of the OKL was irreversible. The remaining aircraft were finally assigned to the Stuka training units where there were eventually worn out. Nobody considered their possible future use as anti-tank aircraft. During the war, the Hs 123 was just as combat ready over Poland, France and the Balkans as during the difficult winter of 1941/42. In the opinion of General von Richthofen, 1000 of these ground attack aircraft could well have change the course of the campaign in the East.

Because production facilities had to be re-tooled in the meantime, mass production had to come to a halt.

The HS 123's greatest asset lay in its robust construction and the resulting ability of the fuselage to remain intact.

The Ha 137, which was created by the Hamburger Flugzeugbau as a competing model and was proposed by the RLM in 1934, had no realistic chance of going into mass production over the Hs 123. On October 9, 1934, a viewing of the mock-up took place at Blohm & Voss. Seven months later, on May 13, 1935, an Ha 137 V1 took off on its maiden flight powered by a BMW 132 radial engine. Then came the V2 model, which also had the BMW radial engine, which was followed by the Ha 137 V3 equipped with a Rolls-Royce "Kestrel" engine, as well as two other test models which were equipped with the Jumo 210 A-Reihenmotor in-line engine. Although Richthofen voiced his favor for the high-performance aircraft, which could also be utilized against lightly armored targets, he was not able to convince Udet. The Ha 137 V4 and V5 aircraft served for a long time, however, in Rechlin as test beds for, among other things, tests with small air-launched rockets against armored ground targets.

Below: the HS 123

The Henschel Hs 123 represented the first dive-bomber built especially for the Luftwaffe. While the A-series was outfitted with the Junkers Ju-Pak two-blade all-metal propeller, a few of the earlier V-models were equipped with adjustable VDM three-blade propellers and a strong BMW 132 K engine. These aircraft were to serve as prototypes for the B and C series, which were abandoned in favor of the Ju 87.

Below: The Hs 123 also took part in the Flug meeting air festival in Zürich-Dübendorf in 1937.

Left: The Hs 123 V2, registration D-ILUA, already had the NACA-type engine cowling, as did the V3 and V4.

Right: An Hs 123 A-1 with the "WL" indicator typical of Luftwaffe training aircraft in 1939. It's complete tail number was WL-ELH. The lack of the Balkan Cross indicates an aircraft which was not equipped for military action.

Left: Beginning on 25 October 1939, training units received the new "S" designator. This is an Hs 123 A belonging to a flight school in the Luftkreiskommando III (Dresden) area.

An Hs 123 A-1, armed with two MG 17 machine guns in the fuselage and four ETC 50's slung beneath the wings. Instead of bombs, this Stuka could be fitted with two MG FF in the munitions canister for use as a fighter-bomber in the close-air combat units on the Eastern Front. A droppable fuel tank is seen under the fuselage.

Below: Aircraft of Lehrgeschwader 2 during the warm-up phase prior to takeoff. The aircraft were chocked in place during the runup.

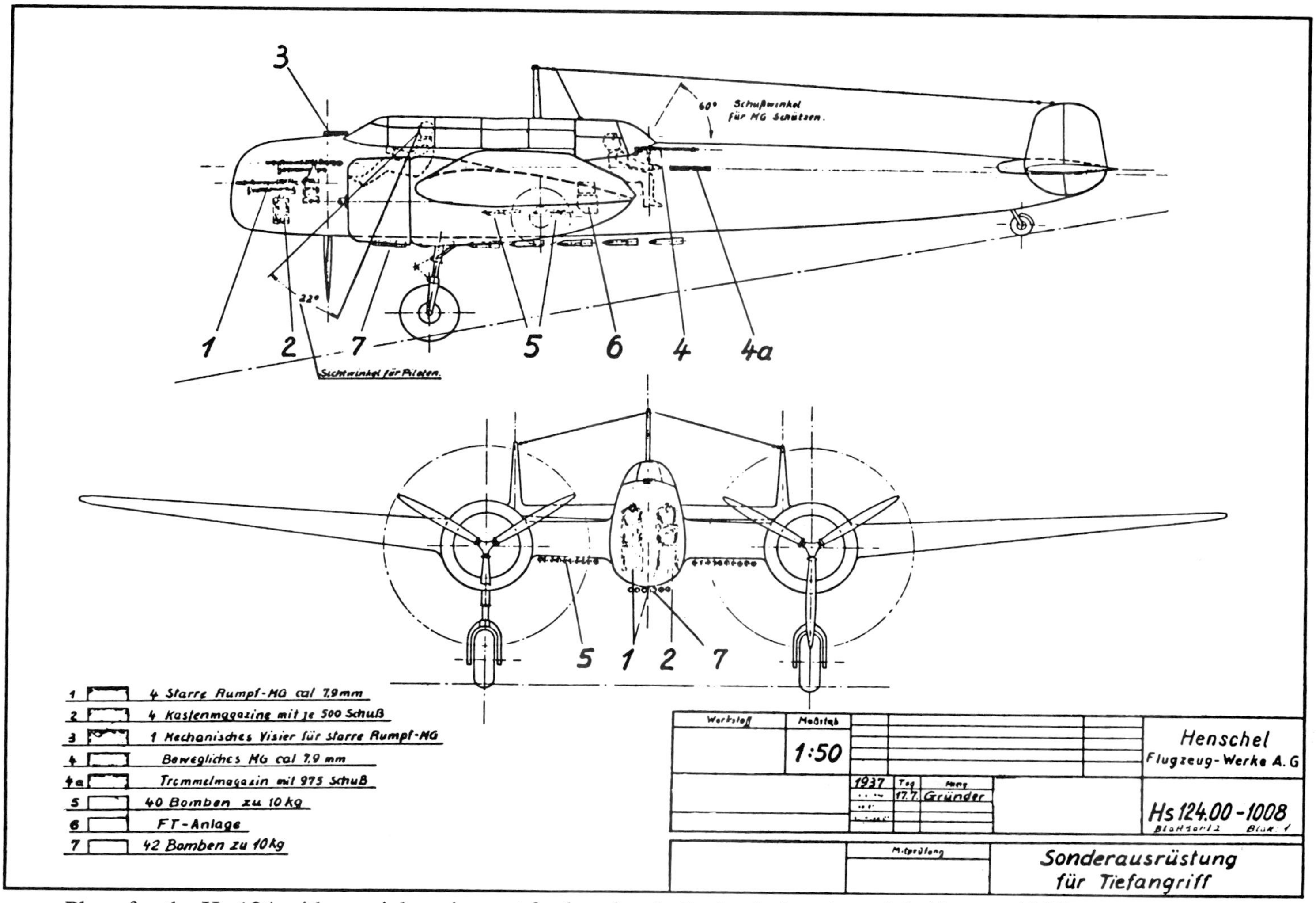

Plans for the Hs 124 with special equipment for low-level attacks dating about July/August 1937.

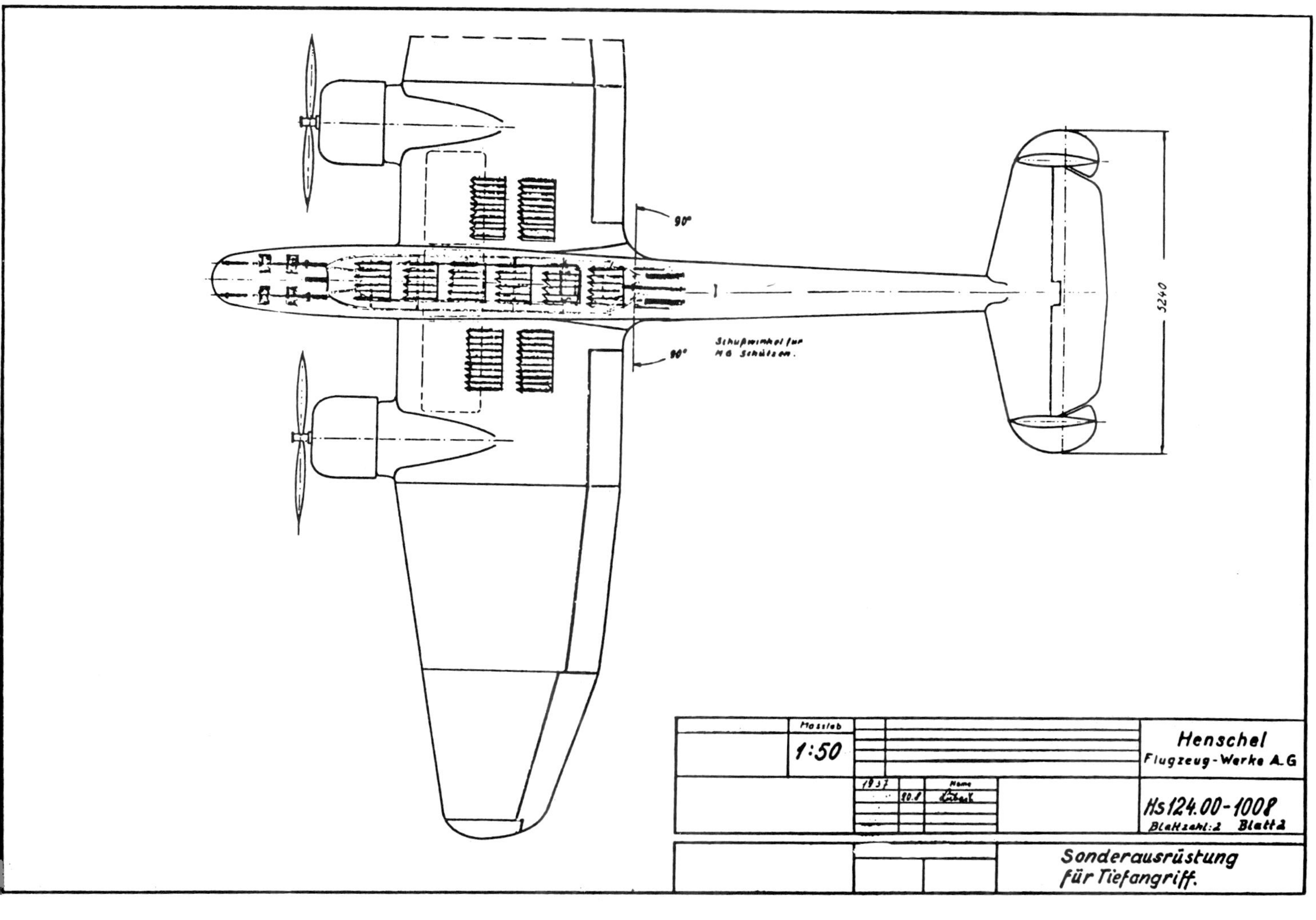

Above: The Henschel Hs 124 was designed to compete against the Fw 57. The individual test versions had different weapons and engines. The Hs 124 V2 pictured here had the BMW 132Dc engine.

Below: Blohm & Voss conceived the Ha 137 in response to the dive-bomber development contract. Dr. Voigt supervised the construction. For his design, he drew mostly from the last aircraft he developed for Kawasaki in Japan, the C.5 air mail carrier. What resulted was a single-seat dive-bomber which was of low gull-wing monoplane design. The engine was a Jumo 210 C twelve-cylinder in-line engine, as it was with the Arado Ar 81. The fixed armament, consisting of two 20mm MG FF and two MG 17 machine guns was quite refined. Plans called for the installation of one 250kg bomb under the fuselage and four 50kg bombs under the wings as additional combat load. In reality, it was clear that a load of only 200kg was possible.

Above: The Ha 137 from another viewpoint.

Below: The first prototype of the He 118 with its new D-UKYM registration. The registration had been D-IKYM. It had to be changed because once it had been modified, it belonged to a new performance class.

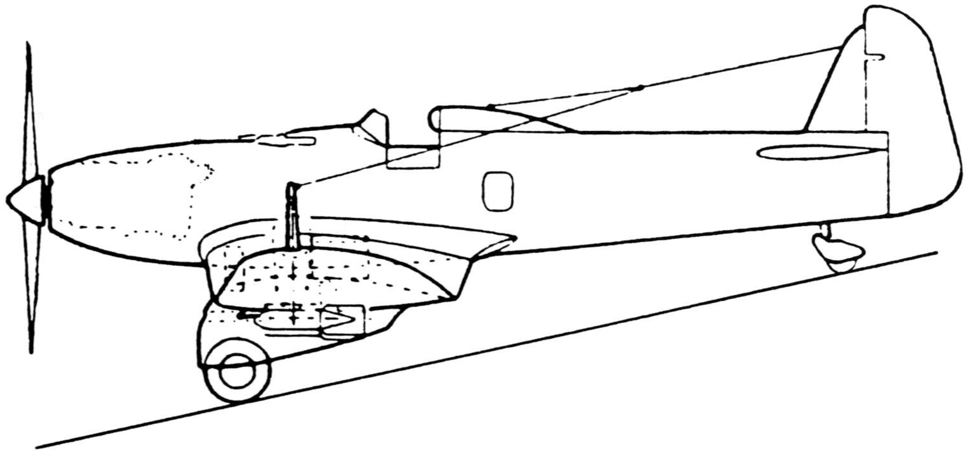

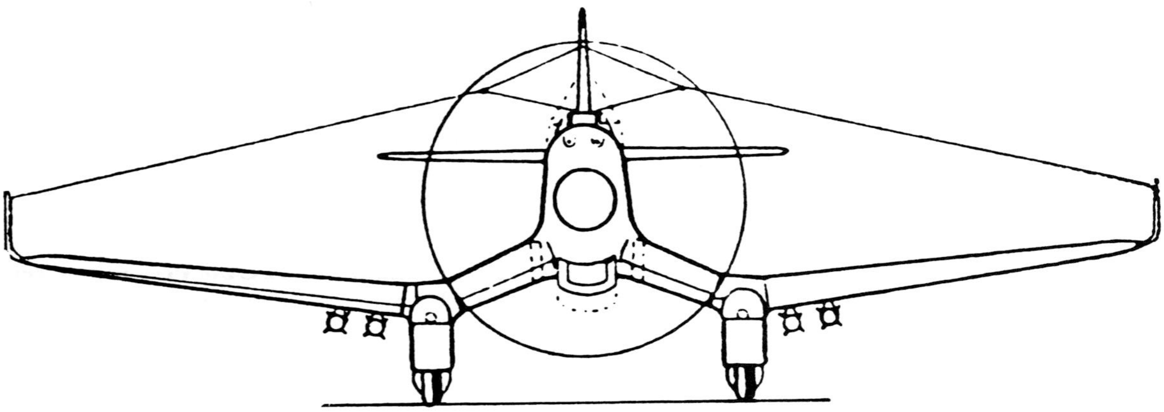

Three-sided view of the Blohm & Voss Ha 137 B.

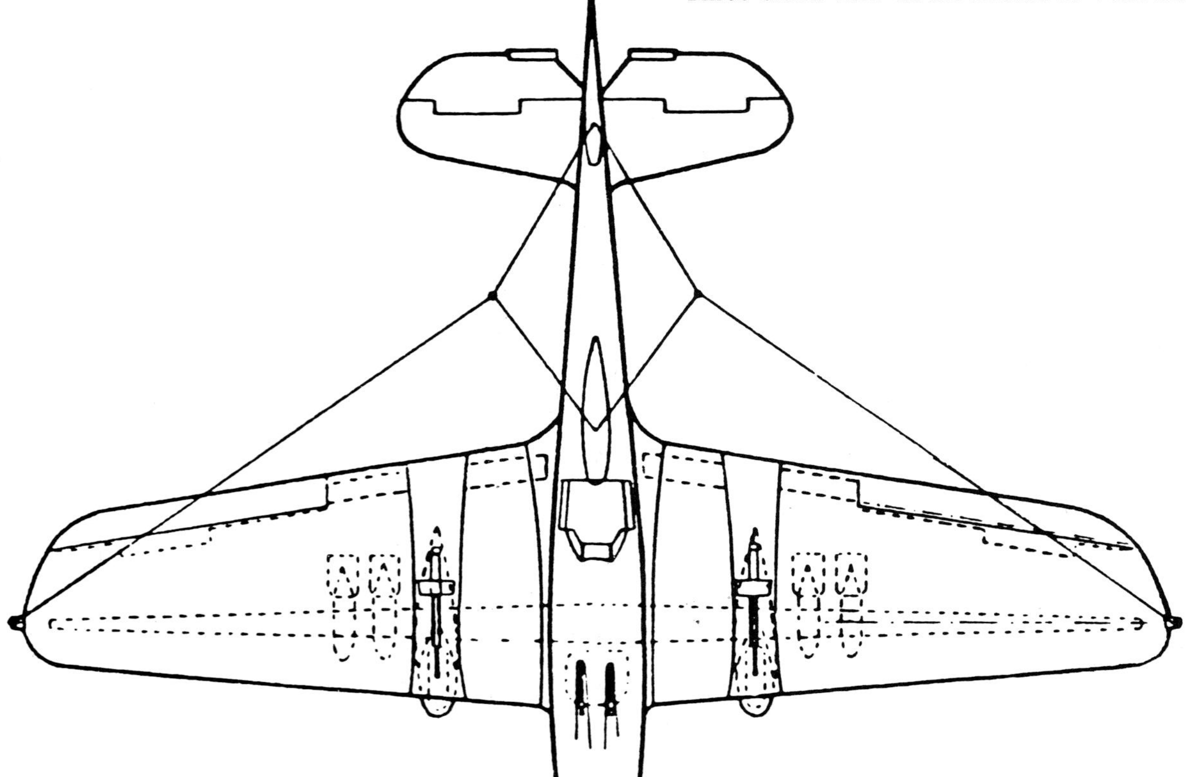

Hs 124, He 118 and Fw 189

The Hs 124 V4 was also developed as a two-seat dive-attack aircraft beginning in 1937. The V1 through V3 models served as a medium bomber concept aircraft. The test model underwent comprehensive statistical testing beginning in January 1938 which showed the necessity for reinforcing the wing. On May 15, 1938, the Hs 124 completed its maiden flight as a dive-bomber, however the engines had to be changed a few days later at Schönefeld. Because some unnerving oscillations occurred during flights conducted in the following months, the wings were again stiffened. New production delays were encountered as a result of a crash landing on November 18, 1938. On February 23, 1938, the RLM permanently grounded the aircraft. The Hs 124 V4 was turned over to E-Stelle Travermünde in June of 1936 for further testing. The whereabouts of the dive-bomber equipped with 20mm cannon is unknown.

During this timeframe, the Heinkel factory was working intensively with the He 118, a medium dive-bomber.

The Heinkel He 118 saw its roots in the Heinkel P-1030 project and was created based on the same RLM announcement which solicited the Ar 81, the Ha 137 and the legendary Ju 87. On February 14, 1936, the He 118 V1 (serial number 1293, D-UKYM) made its first flight. Nitschke, an experienced factory aviator, was its pilot. As opposed to the V1, which was powered by a Rolls-Royce "Buzzard" engine, the He 118 V2 had a Daimler-Benz DB 600 C in-line engine. Another three prototypes followed these two aircraft, as well as a "zero" series consisting of eight aircraft, three of which had DB 601 engines. Most of these aircraft were armed with an ETC 500 and an MG 15 machine gun for the observer. Heinkel armed the "zero" series with two additional wing-mounted machine guns. The cockpits of these models were partly open, partly closed. On December 15, 1937 the first aircraft of the zero series, the A-01 with dive brake, reached the amazing speed of 595 km/h during a dive.

The Focke-Wulf Fw 189 was not conceived as a fighter-bomber to engage armored or soft ground targets, however some test aircraft did receive a heavily armored cockpit as well as offensive weapons consisting of six fix-mounted guns. The two seater, designated the Fw 189 V1a and V1b, later received an observer's position which was altered on numerous occasions. This aircraft was not mass-produced, but because it could carry up to four SC 50 bombs under the ETC 50/VIII, it was well suited in knocking out light vehicles even with a near-miss. These aircraft primarily flew low-altitude reconnaissance missions.

The Heinkel He 118 was the most advanced design in response to RLM's contract invitation.

The He 118 was captivating not only in its attractive aerodynamic exterior form, but also had, as opposed to the other competing models, retractable landing gear which helped give it superior performance. It met the requirements for bomb load and armament. Nevertheless, the decision fell in favor of the Ju 87. The reasoning for this was that the He 118 construction had not been entirely completed by the time of the comparison flight testing and could therefore not enter a vertical dive. Also, Ernst Udet had an accident with the He 118 during later phases of flight testing on June 27 1936. While in a dive, the engine accelerated past redline, the propeller flew off and Udet was forced to parachute to safety.

Below: In 1940, attempts were made to convert the Fw 189 reconnaissance aircraft into a heavily armed ground attack aircraft by installing a two-seat armored cockpit. In mass production it was to be called the Fw 189 C. Aircraft D-OPVN was converted from the V1 series to the V1b by changes to the cockpit. There also existed a V6 prototype for the C-series.

Single-engine Fighter-bombers

Bf 109

It was only a question of time if the Bf 109 would also be equipped with a bombing system. With such a system, the E-3/B and E-4/b could each carry up to four SC 50 bombs. From the Bf 109 F-1 to the F-4, there was a fighter-bomber version of each equipped with an ETC-250 bomb rack. Later, the R6 bomb rack was introduced which made the utilization of 250 kilogram bomb load possible. The G model could also be used as a fighter-bomber against armored and soft targets, as the Fw 190 came to be a part of the inventory of ground attack squadrons in ever-increasing numbers beginning in 1943. The Bf 109 G-2/R1 was also tested in this role, and was fitted with a third oleo strut so that the 500 kilogram bomb load would not touch the ground. Two 300-liter droppable fuel tanks under the wings served to increase the aircraft's range. The Bf 109 G-3's landing strut mounted bomb racks could accommodate either an ETC 500 IXb or four ETC 50 VII-ld. The G-4 and G-6 could also be fitted in this manner. In the case of the Bf 109 G-10 and G-14, there was an R-I system which turned the fighter into a fast fighter bomber. Beginning in 1944, fighter squadrons, particularly those on the Eastern Front, were used in low-level engagements. In this manner, pilots of JG 11 were able to destroy 186 trucks and disable another 161 in 713 low-level sorties in February of 1945. Eight tanker trucks we set afire, 250 horse and carriage teams and 30 artillery guns were destroyed. Three light tanks were rendered motionless. The engagement of heavy tanks was to be left to Ju 87's as well as Hs 127's.

The next series, the Bf 109 K-4, could be armed with either 250 or 500 kilogram bombs. In comparison with the Fw 190, this model of the Bf 109 was hardly used as a fighter-bomber or in the anti-armor role.

Fw 190 F and G

After a bomb rack was fitted to the Fw 190 A for carrying various droppable loads, similar to the Bf 109, the construction series F-8/R1 to R3, F-9 and F-15 as well as four models of the ground attack version F2 190 G were conceived. This version were to fulfill a special role in the defensive fighting against the Soviet tank regiments on the Eastern Front.

The Fw 190 F-8/R1, which was driven by a BMW 801 radial engine, had one fuselage-mounted ETC 501 and four wing-mounted ETC-50's. Its fixed guns were two MG 17 and two MG 151/120 machine guns. The R2 varied from the R1 in its fuel tank system in the fuselage. The F-8/R3 was only produced in limited numbers and had an MK 37 103 cannon under each wing, each having 35 rounds. This particular weapon proved to be quite effective on the well-armored Hs 129, an aircraft well-suited to engaging enemy armor.

The F-9/R1 was identical to the F-8 but had the stronger BMW 801 TS or TH engine. The following series were designated as F-15 and were given a beefed-up landing gear and an ETC 504. They were often deployed in defense against enemy armor attacks.

While the G-3 was planned as a quick strike aircraft with two 300 liter droppable tanks, the G-3/R5 was quite successful as a ground attack aircraft.

As an example, on July 8 1943 about 60 T-34 tanks were reported by the commander of the IV./ SG 9, against which Major Druschel's Fw 190 equipped with SD 50 bombs and an Hs 129 were deployed. Reports indicate 50 destroyed or immobilized Soviet tanks.

By converting the F-8 aircraft into G-8's and by retrofitting them with two ETC 71's instead of the four ETC 50's, the series Fw 190 G-8/R5 was created. This ground attack aircraft had two MG 151/20 machine guns in the wing roots as its fixed weapons which permitted attacks on lightly armored vehicles. Most of these aircraft were with Schlachtgeschwadern SG 4 and 5, with Nachtschlagsgruppe 20 and III./KG 200.

Right: The Messerschmitt Bf 109 also saw use as a fighter-bomber. By adding the ETC 250 bomb rack, droppable munitions of up to 250kg could be delivered to the enemy. Here is an aircraft belonging to I./ JG 54, better know as the Grünherzgeschwader, in typical winter camouflage.

Above: A Bf 109 E-4/B of II./JG 54.

Below: When installing the SD(Top) 500 on the BF 109 F, a support wheel had to positioned under the fuselage in order to prevent the external load from touching the ground during takeoff.

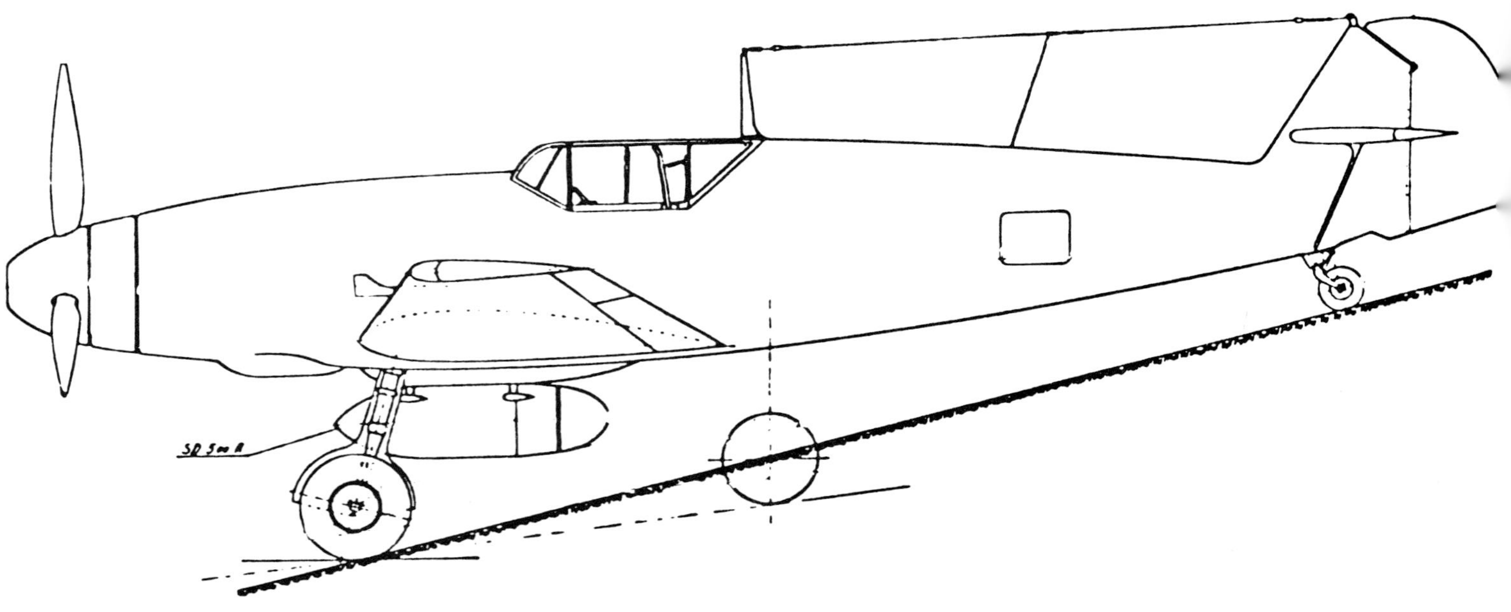

Left: A Focke-Wulf 190 A-3 of Sch.G. 1 returning from a mission.

Right: Four SC 50's under an early version of the Focke-Wulf 190 A during testing of droppable munitions.

Left: This fighter-bomber belonging to JG 5 in Finland during the winter of 1943/44 is being fitted with bombs for its next mission. It's interesting to see the sleds being used to transport the bombs in the snow.

Above and left: Slinging the SC 250 from the ETC.

Above right and below: As there were insufficient numbers of the F-series ground attack aircraft with the wing-mounted ETC, A-series fighters were given diminished wing guns and the ER 4 bomb cage which could hold four SC 50 bombs was mounted under the fuselage.

The Standard - the Ju 87

The construction of the first mass-produced series of the Ju 87 began in November 1934. Shortly after the first flight on September 17, 1935, the V2 model flew for the fist time on February 25 136. Eleven prototypes followed up to May 1939. Comparative flights conducted with the Vought 156 and the Curtis 77 clearly showed the advantages of the Ju 87. The Ju 87A went into mass production beginning in June of 1937.

The Junkers factory then started work on the development of improved Ju 87 models. The fist prototype, the Ju 87 V7 (serial number 4926) flew on November 11, 1937 and received good reviews.

By this time, the first Ju 87's had been deployed in Spain and had fueled the argument for dive-bombing. At the same time, these aircraft were tested as a means of defense against armor.

As opposed to the Ju 87 A-1, which was powered by a Jumo 210 D, the B-1 received the more powerful Jumo 211A in-line engine. Its weaponry was like that of its predecessor; two MG 17's and one MG 15 machine guns. With the B-2 model, the bombload could be increased from 500 to 1000 kilograms. The un-armored version of this series of aircraft was allowed to quickly run out. Similar in construction to the B-series were the Ju 87 R-1, R-2 and R-4 dive-bomber aircraft. These aircraft differed in their ability to carry two 780 liter external fuel tanks as well as bomb racks in the central wing area.

The V 22 and 24 test beds were related to the D series as its predecessors. The V 22 model of the Ju 87 crashed on August 22, 1942 in Rechlin. For this reason, the V 24 (serial number 0544) was transferred there and thoroughly tested.

The aircraft had a Jumo 211J engine, a new cockpit shape, enlarged wings and much improved equipment. During the introduction of the D-1 series, eight aircraft were lost in Southern Germany when they made emergency landings due to technical malfunctions. Based on this, the RLM called for numerous alterations to construction at the end of 1941, among which were improved armor, simplification of the bombing system as well as the installation of two 20mm MG 151 machine guns in the wings. The first test aircraft of such a low-altitude/ground attack aircraft (D-2), which was serial number 2013, was handed over to Tarnewitz for weapons testing in March of 1942.

During this testing and that done at the factory test facility, problems with the cooling system had to be dealt with. These aircraft and the five D-series which followed proved to be of particular value in defensive action against strong armor and infantry units.

Ju 87 development was broken off in early 1943 and Junkers converted earlier models to training aircraft (construction series H-1 to H-8).

Aside from the Junkers factory itself, the Weserflug factory was the most important licensed manufacturer of Ju 87's. The first aircraft built there conducted its maiden flight on February 26 1938. The 1000th Stuka to be produced by Weserflug had already left the assembly lines by February 27, 1941. Based on the tactical air situation, the production had to be moved in 1944. Numerous aircraft, for example, had to be produced in the basement corridors of Berlin-Tempelhof airport, where later the final assembly of Fw 190's was also carried out. In September 1944, the production of the Ju 87 was halted in favor of the introduction of faster fighter-bombers.

The first ten Junkers Ju 87 A-0 left the assembly halls in Dessau early in 1937 and formed the nucleus of I./Stukageschwader 162 "Immelmann." These were later joined by the Ju 87 A-1, which conformed with the "zero" series. However, the Ju 87 A's performance, with a top unladen speed of 320 km/h, was not sufficient. Even its bomb load of 500kg could be carried only without a radioman and defense weapons on board.

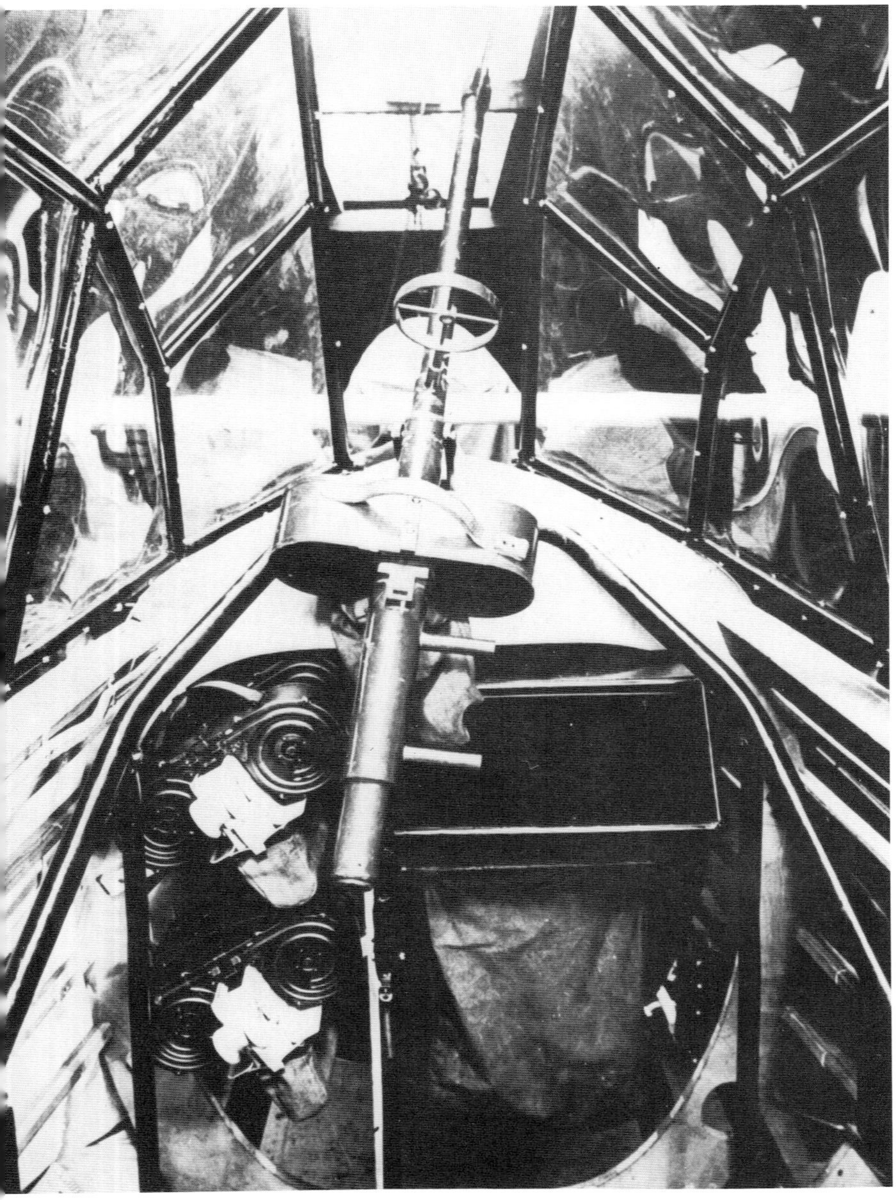

Left: The MG 15 (7.92mm caliber) formed the Ju 87 A's defensive weapon. The supply of replacement magazines can be seen in the lower left of the picture

Below: A Ju 87 A with an ETC 50 under the wings was a research test for the B-series, which received four more ECTC 50's under the wings in addition to that under the fuselage.

Above: “Aviators’ Memorial” of an early version of the Ju 87 B-1. The installation of its Jumo 211 A with 1,200 horsepower yielded a definite improvement in performance. Its unladen speed increased to 350 km/h and the combat load increased to 1,000kg. Operations over enemy territory were, however, only possible after air superiority had been achieved. This was to become quite apparent during the Battle of Britain.

Below: A Ju 87 B of StG 51.

Above: The NSU Kettenrad tracked motorcycle was an outstanding transport and towing vehicle in the difficult terrain of the Eastern Front. In the background is a Ju 87 D equipped with skis.

Below: This Ju 87 B served in preliminary research for the installation of the Jumo 213 E. For this purpose, the Bär firm built a special radiator which was tested in this aircraft.

Left: This device was used to lift droppable munitions into their bomb racks.

Right: The weapons canister known as WB 81 A which held three MG 81A machine guns and 250 rounds of ammunition was used for attacks on non-armored targets.

Left: The installation of the Jumo 211 J with 1,300hp, the flatter radiator and a newly constructed cockpit canopy gave the standard dive-bomber an entirely new look. Its top speed could be increased, albeit only slightly, and the combat load upped to 1,600kg. The D-series was equipped with the MG 81Z machine gun as a defense weapon. The late-model Ju 87 D-3 seen here had supplemental armor as well.

Left: Bombs slung from this late model Ju 87 D-1 are fitted with proximity fuses. These permitted the bombs to explode above the earth's surface.

Below: This Oberfeldwebel (foreground) and Unteroffizier beam for the camera with obvious pride in a Ju 87. Both have been awarded the EK 1 (Iron Cross) and wear the pilot's badge.

The "Kanonenvögel" and Destroyer Aircraft

Ju 87

The Ju 87 G-2, the only plane other than the Hs 129 to be deployed in appreciable numbers as a standard anti-armor aircraft, had its roots in the G-1 series and was armed with two BK 3.7 cannon. The first 25 of these specially-armed aircraft were delivered up to March of 1943. A little later, the Luftwaffe ordered an additional 22 Ju 87 G-2's. After their initial testing phase, these aircraft were stationed with Panzerjägerstaffeln 1 and 2.

In an detailed meeting on August 1, 1943, the future of anti-armor operations on the Eastern Front was the topic of discussion. Therefore, a high priority was given to the destruction of enemy tanks because the heavy Pak's (Panzerabwehrkanone) could not always be in position on time due to unfavorable influences of weather and poor road conditions. While the aircraft's testing showed that the Ju 87 G could also only be considered an intermediate solution because they were seen by the pilots as not maneuverable enough and too susceptible to enemy fire. This resulted in many of these flying "Panzerknacker's" receiving additional armor to give the crew added protection. The last missions were flown in March and April 1945 by Ju 87 G-2's of 2.(Pz)/SG 2, 10.(Pz)/SG9 and 10.(Pz)/SG 77. As a result of the massed anti-aircraft artillery of the Red Army, these missions resulted in heavy losses.

For engaging either non-armor targets or troop concentrations, the WB 81A (15 degree depression angle) or the WB 81B (0 degree depression angle) gun pods could be used instead of the BK 3.7. These pods were slung from each wing's hardpoints together with ejection devices for the linked ammunition (belts of 100/500 rounds). These belts had mostly SmK-, SmL- and PmK ammunition linked in a 2:2:2 configuration. The use of these pods was somewhat limited as the MG 81 was urgently needed as a defense weapon for almost all German aircraft.

Above: In December 1942, a Ju 87 D-1 was armed with two 3.7cm Flak 18 anti-aircraft guns. After the system was proven to be fit, other aircraft, in particular the Ju 87 D-3, were armed with the system in February 1943. The nomenclature of the altered aircraft was then G-1. In the Ju 87 G version, there were no speed brakes and also no possibility of carrying any self-defense weapon.

Above and left: The BK 3.7 was developed from the Flak 18 by altering the recoil mechanism, shortening its run, and installation of an ejection system. After modification, the barrel length of the cannon was 120.60cm, its overall length 375.00cm. With the hardened core ammunition tested at the E-6 Gruppe in Reichlin beginning in February, the gun had a penetration capability of 58mm at an impact angle of 60 degrees at a distance of 100 meters. This penetration capability was increased to 120mm when the Pz.Gr. L'spur was utilized later.

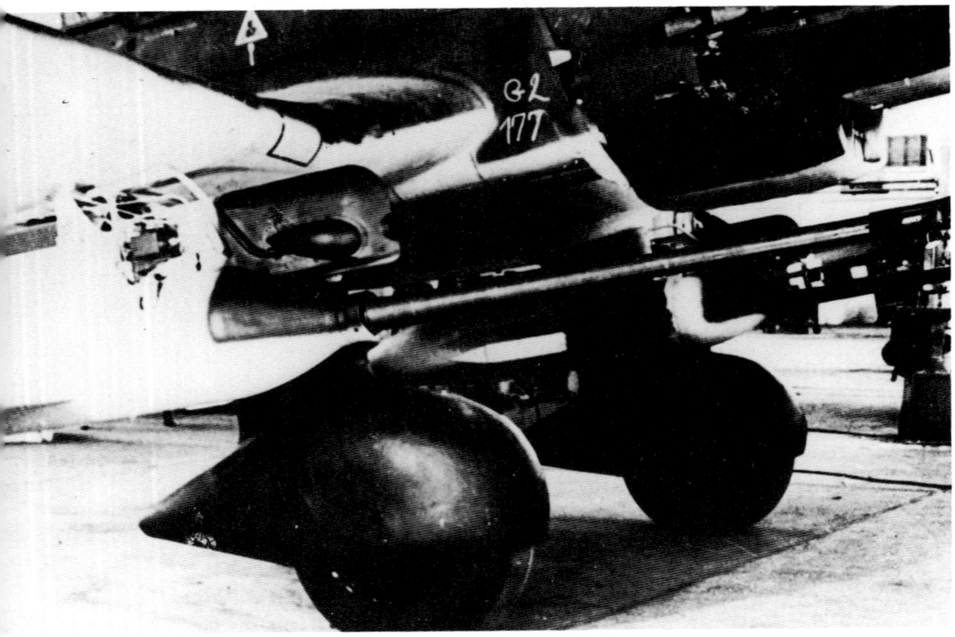

Above: A Ju 87 G-1 of Panzerjagdkommando Weiss in Russia. Each Flak 18 had twelve rounds. There were six Wolfram grenades in one magazine on the side of each weapon. The tip of the spinner is painted red.

Left: The Junkers Ju 87 G-2 was a D-5 retrofitted with a BK 3.7 and a longer wingspan for anti-armor assignments. About 208 aircraft were rebuilt in this fashion.

Left: When zeroing the weapons at the range, the Ju 87 G-2 was put on blocks and aligned.

Ju 88 P

In addition to the Ju 87 G-1/G-2 models with their 3.7cm rockets, Junkers developed four more basic types of heavy anti-armor aircraft based on the Ju 88 A-4 bomber. The first trial installation was conducted on a Ju 88 V52 (serial number 2047) and consisted of a 7.5cm anti-armor gun.

The prototype of what was to later be known as the P-1 series received a fortified gun carriage and test-fired 240 rounds into steel plates and a captured tank. The results of this firing indicated necessary construction improvements to strengthen the fuselage nose section and its planking.

In about March of 1943 an additional twelve prototypes were converted under contract. Plans called for building up another seventeen P-1 type aircraft. Later, 30 of these anti-armor aircraft were to leave the assembly line per month.

Because the manufacture of the heavy anti-armor Pak 40 (BK 7.5) was so inadequate, this plan could not come to fruition. The Oberkommando des Heeres (OKH) therefore suggested the installation of the armored personnel carrier gun KWK 39 (what the Luftwaffe called the Bordkanone BK 5), which was somewhat smaller in caliber.

After installation research revealed no misgivings, the first contract for 100 Ju 88 P-1's was quickly canceled. Nonetheless, a few P-1's were tested by the Einsatzkommando Ju 88 (PrJ) in field conditions in the East in the summer of 1943. Due to the fact that the aircraft could not sustain hits from enemy fire very well, the OKH voiced its support a few weeks later not only for the Hs 129 B-3 but the Ju 87 G-1/G-2. This despite the evidence that a 7.5cm round was sufficient to knock out enemy armor at the time.

The planned conversion of III./KG 1 to Ju 88 P's, which had already begun, was broken off early.

Various muzzle brakes were used for the prototypes of the BK 7.5 beneath the Ju 88 P-1.

The first prototype of the Ju 88 P-2, serial number 140452, was also called the Ju 88 V57/1. This aircraft was fitted with two BK 3.7's in a roomy floor turret. In July 1943, the first steps were taken in the direction of the P-4 model with the strengthening of this floor turret. Although this weapons arrangement had proven itself effective, the RLM canceled this series. The P-3 which followed was just like the P-2, but the aircraft was to receive improved armor in the mass-produced version. Because this improved armor increased the weight of the aircraft, at the cost of flight performance, this version remained a prototype. After cancellation of the first three construction models after only a few prototypes and "0" series models, Junkers returned to the V57. In addition to a Ju 88 S-1 with two KWK 39's in a newly created floor turret, the V57 was armed with a 50mm gun. After a test firing of 50 rounds –conducted using semi-automatic loading device – the aircraft was transferred from Tarnewitz to Dessau where it was overhauled. Although a mass-production of these aircraft was out of the question at the end of 1943 due the capacity of the industry, the P-4 model was the only one, however, to be accepted into the inventory by the RLM.

In addition, there were a few prototypes with MK 112 cannon, the Ju 88 V53 (serial number 1182) as well as at least on prototype with the Düsenkanone (Düka) 88. This prototype was completed by the end of 1944 and given considerable testing. It did not see action against enemy armor. This was also true for the tests which were to outfit a Ju 88 P-4 with a battery of smoke canisters to carry out the so-called "Eisenbahnbekämpfung."

Above left: Because the Pak 40 was not readily available, the OKH suggested using the KWK 39 tank gun, with a caliber of 5cm, which was called Bordkanone BK 5 by the Luftwaffe. This model was called the Ju 88 P-4.

Conversion to the Ju 88 P-1 was accomplished in July 1942. The 7.5cm Pak 40 was utilized here. The aircraft shown here was part of Panzerjagdkommando Ju 88, where it flew operational sorties in the summer of 1943.

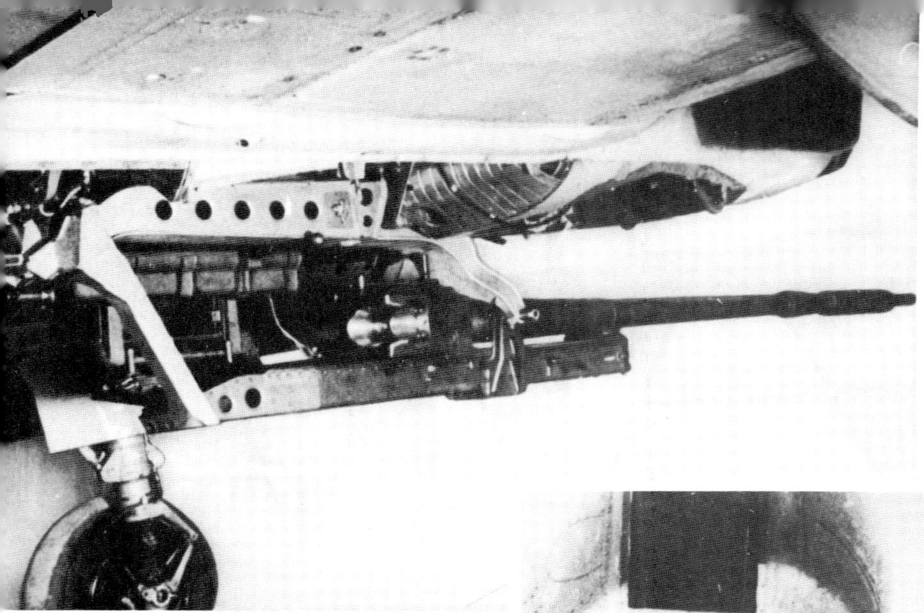

Left: Well seen in this photo is the gun carriage of the 7.5cm cannon and how it was mounted under the fuselage of the Ju 88 P-1. The flooring was removed in this case.

Right: Attempts were made to use the 7.5cm cannon on the aircraft in order to also be able to engage heavily-armored targets. The Ju 88 A-4 was well suited to this, and with the BK 7.5, it was called the Ju 88 P-1.

Left and below: The Junkers Ju 88 P-2 received two BK 3.7's in a newly designed roomy belly casing. The retrofitting of this Ju 88 V57/1, serial number 140452, was accomplished in July 1943.

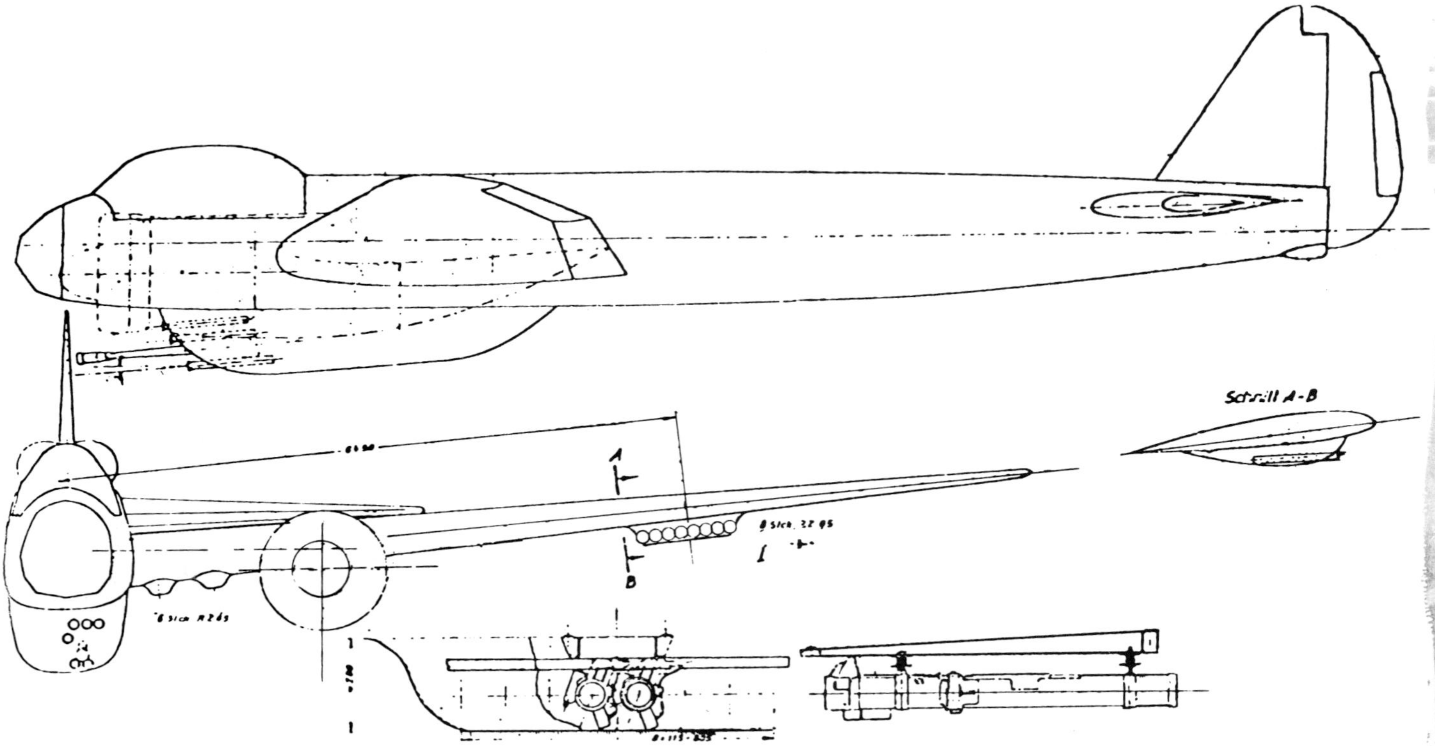

Above: Overview sketch for the installation of RZ 65 gun barrels in the Ju 88 P-4.

Below: Installation of the Bordkanone BK 7.5 in the Ju 88 P-1.

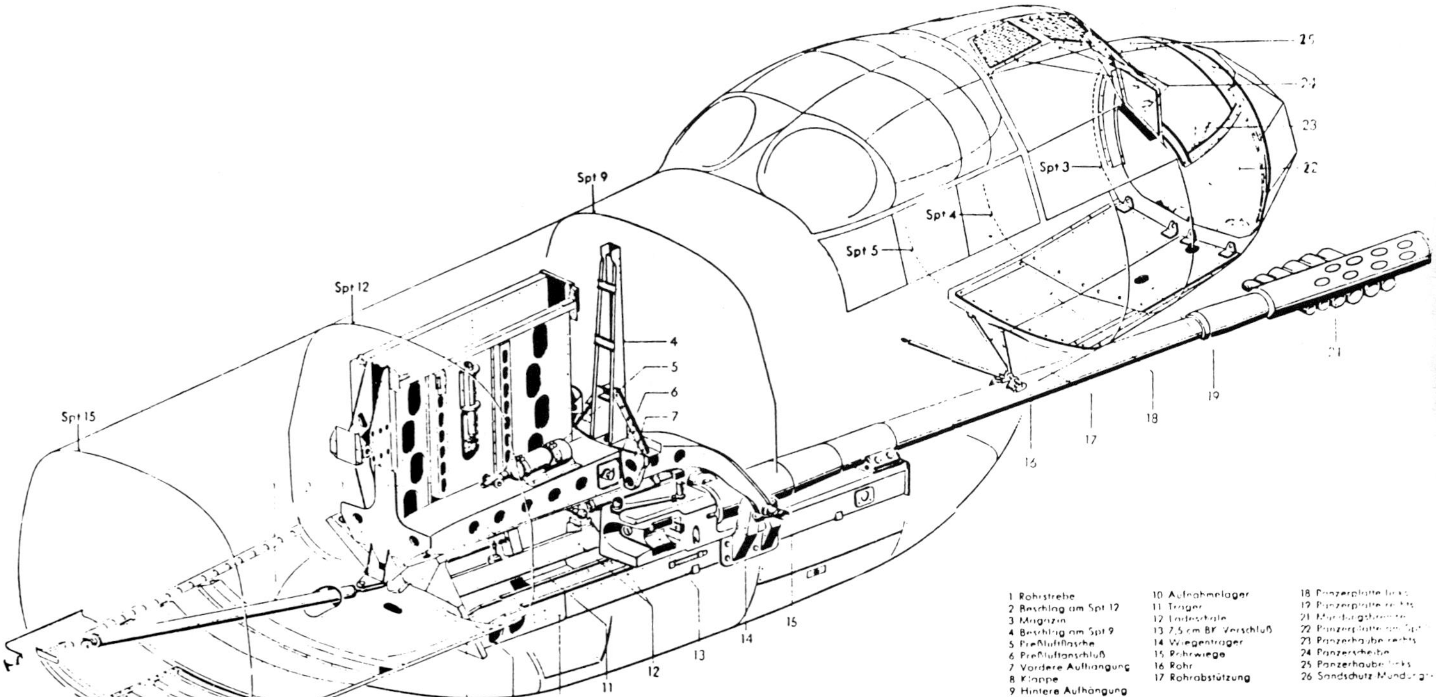

The Ju 88 V57/1 received the designation of Ju 88 P-2 after the installation of the BK 3.7.

Hs 129

The Heinkel works prepared project 26, the predecessor to the Hs 129, as a separate model beginning in April 1938. After the first dummy had been built in May, it was ordered about two months later to change the configuration to accommodate two As 410 engines, as those previously specified were no longer available. In November, the construction of the Hs 129 V1 was begun. Because the Argus engines had not arrived, flight testing had to be delayed until May 26, 1939. This was then followed by numerous alterations made to the aircraft.

The second Hs 129 flew for the first time on November 30, 1939, the third Hs 129, the V3, also experienced delays related to lack of engines. Therefore it was not until April 2, 1940 that the factory testing began, which were severely impaired by the insufficient engines in the case of all three prototype aircraft. After the Hs 120 V2 was lost in a crash on January 5, 1940, and the V3 had been damaged, only first prototype could be transferred to Rechlin on July 31, 1940 for testing under actual deployment conditions. In July 1940, construction began on sixteen A-1 models at Henschel in addition to the "0" series (A-0). Because the engines necessary for low-altitude attack and anti-armor operations were still missing, the A-1 series were renamed to B-0 and testing was begun in September 1940 for the Gnome & Rhone radial engines. The testbed for this research was the Hs 129 V3, starting on March 3, 1941, which also broke up during testing. Subsequently, testing of the B-0 prototype began in Rechlin.

Though that was the case, the delivery of the mass-produced aircraft came about very slowly. Up to December 31, 1942, only 248 Hs 129's were delivered to the Luftwaffe, of which 41 were lost in a short time, some due to enemy action, some not. In March, 212 of the original 331 aircraft, that's 64%, were destroyed.

After several of RLM's wishes for change were implemented, the cockpit glass was substantially improved and the engine covering and location of the oil coolers were re-worked. A thorough alteration of the airframe was deemed too expensive, so more powerful guns were worked on as an improvement. The standard armament consisted of one Mk 101 (later it was the MK 103 with 30 rounds of ammunition) as well as two MG 151/20 and to MG 17 machine guns. Up to eight 50-kilogram bombs could be carried on board, in which case the

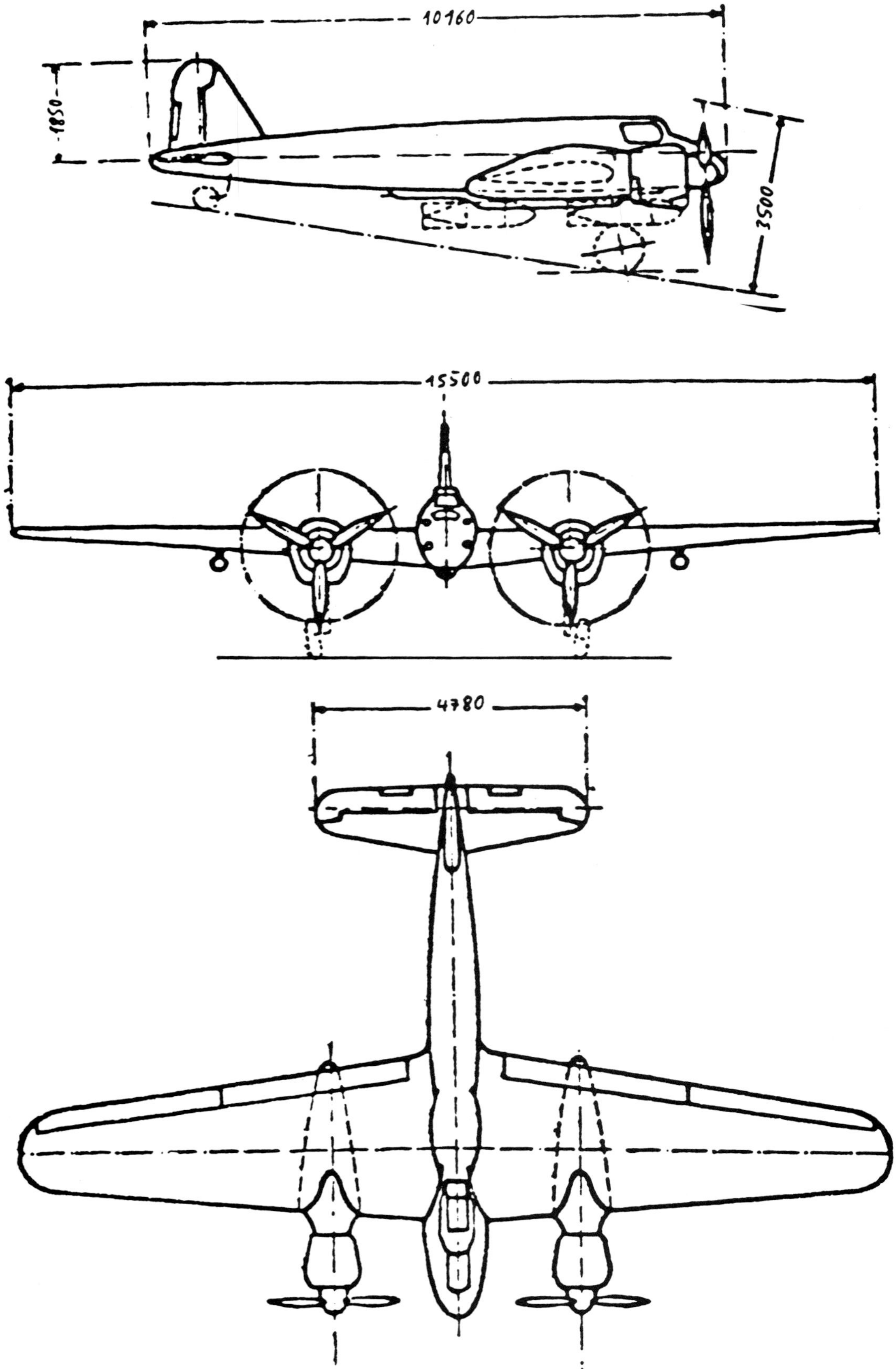

In 1938, the Hs 129 was created from Henschel project HS P 76.

30mm cannon could not be carried aloft. Thanks to the 30mm cannon, numerous victories were achieved on the Eastern Front against enemy armor. The MK 101 did not prove itself to be sufficient in all cases. Beginning in August 1943, four squadrons, the 4. and 8./SG 1 as well as SG 2 and Panzerjägerstaffeln 51 were equipped with the Hs 129.

Later developments encompassed the following improvements and installations:

Left: The Hs 120 came into being at the end of 1938 from the Henschel project P 46 as a twin-engine armored ground attack aircraft. To the left in the photo is a rumanian Hs 129 B.

Right: As the Hs 129 A equipped with the As 410 engine proved to be a poor performer, the aircraft was given Gnome et Rhone 14M engines with 700 horsepower each. This version received the designation of Hs 129 B.

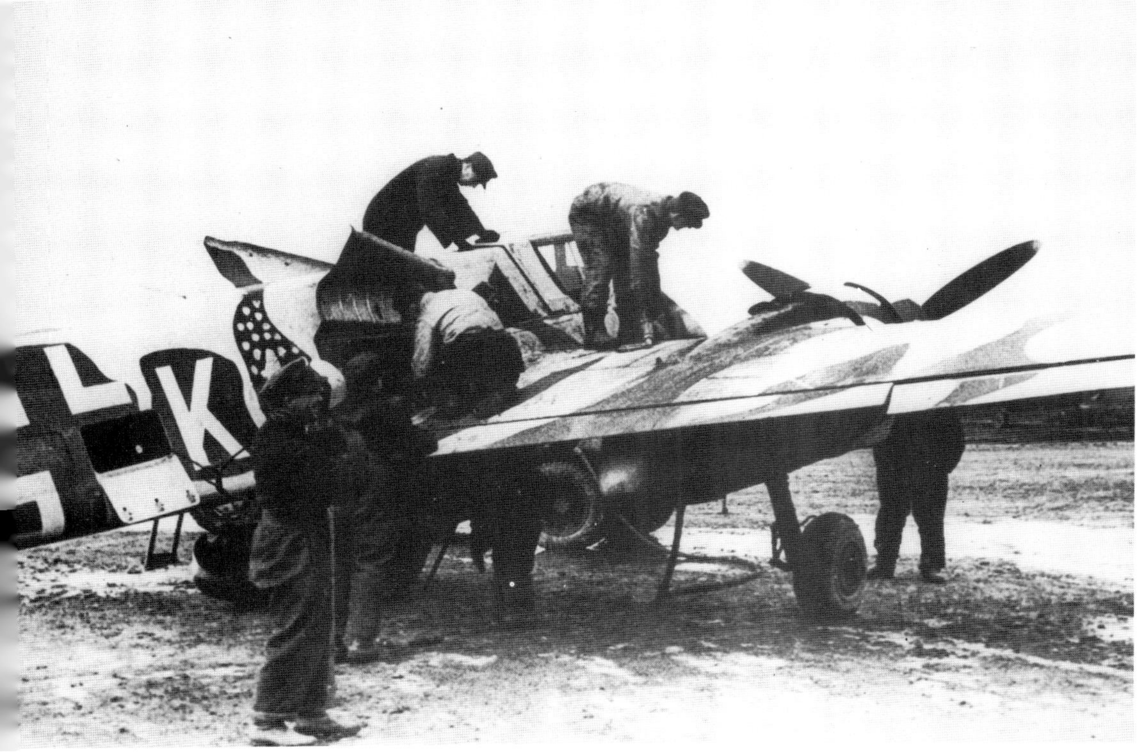

Left: An Hs 129 B-2 of IV./(Pz)SG 9 near Czernovitz in March 1944.

Above and below: Two properly parked Hs 129 B's with protective tarps over the engines and cockpits in the winter camouflage pattern of "white on everything."

Left and below: The first prototype of the Hs 129 B-2 series, serial number 141291, which was equipped with the MK 101 cannon (on the left with cannon system deployed).

Below: Installation sketch for the MK 103 in the Hs 129 B-2.

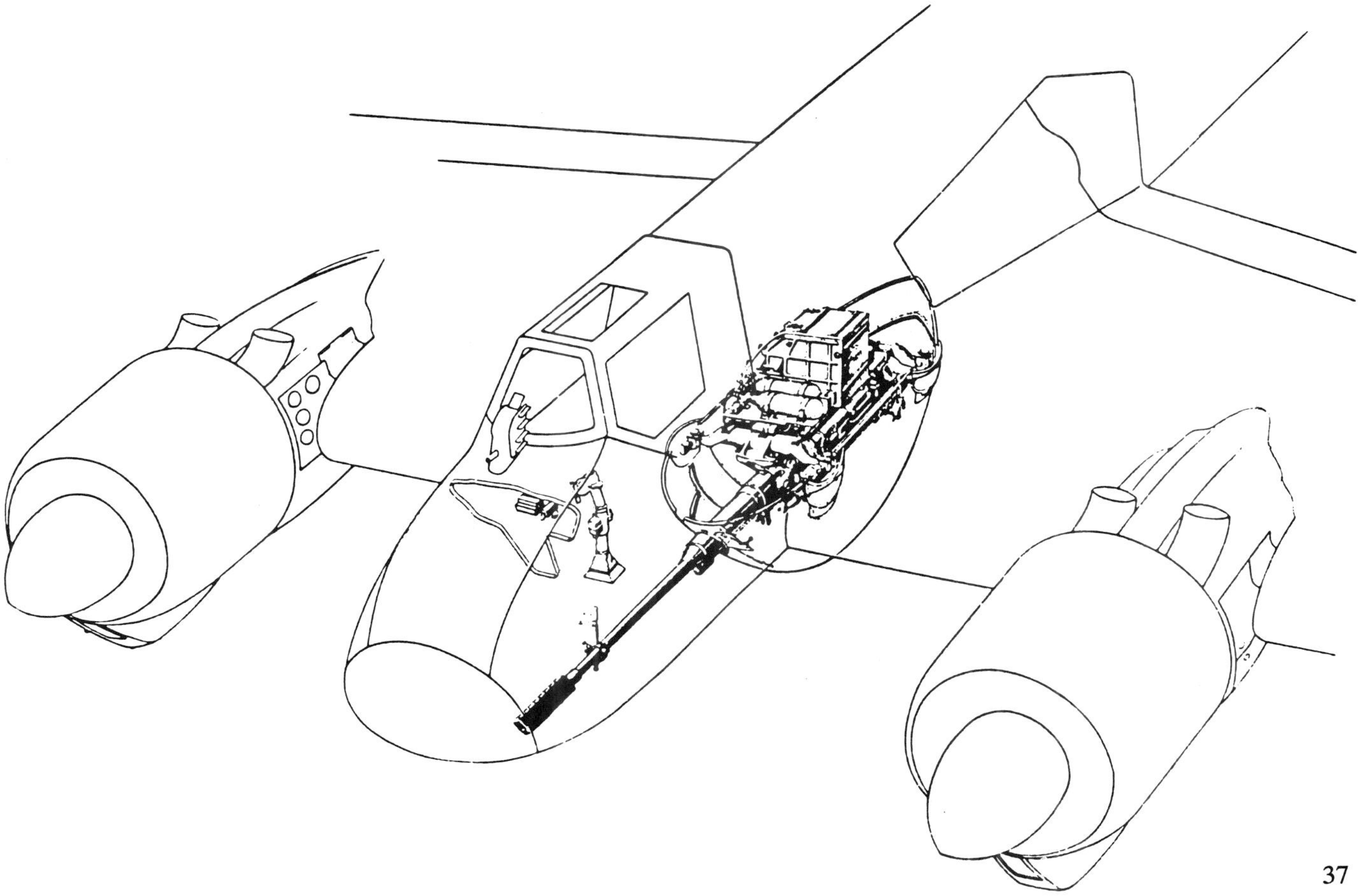

Above: Operational version of the Hs 129 B-2 with the MK 101 cannon. This weapon, which was 30mm caliber, weighed 180kg when assembled. Its overall length was 264.20cm. From a distance of 300 meters, it could penetrate 75mm armor.

a) four MG 17 guns with 1000 rounds (for use on the front)
b) two MG 131 guns with 500 rounds (dummy only)
c) four MG 131 guns with 250 rounds (dummy only)
d) two MG 151's with 250 rounds (dummy only)
e) two MG 103's with 25 round (standard load)
f) 5cm cannon with 12 rounds (dummy only)
g) 7.5cm cannon with 12 rounds (test only)
h) fitting with grate for SD•24 and other bombs (for use on the front)
i) installation of grenade launcher for WGr. 21 and WGr. 30 (test only)
j) installation of special weapon of 5/7cm caliber (test only)
k) installation of a flamethrower with 300 liters (test only)
l) vertically-firing system, automatically triggered (test only)

In the mean time, the B-2 series was created from the B-0 series, which had the modified engine cowlings. The projected C-1 model was to be more heavily armored and have improved protective engine armor.

At the same time, the first prototype aircraft (serial number 140494) were equipped with the BK 7.5 as well as the B-2. The first flight tests were conducted in December of 1943.

Below: This Hs 129 B-2/R-2 (serial number 141859) was operational on the Eastern Front and in the summer of 1943 belonged to 4./Sch.G.1, which became 10.(Pz)./SG 9 a short time later. The additional armament with the MK 103 can be easily seen here.

Left: Flight characteristics of the Hs 129 B-2 were to be tested once this wooden mockup of the BK 7.5 had been installed.

Even in the early part of the year, the deployment of the Hs 129 was carried out – even under very difficult conditions out. OFw (OberFeldwebel) Dittrich was able to destroy 42 enemy tanks during approximately 100 sorties up to April 12, 1943. On March 13, 1944 alone, the Hs 129 managed to knock seven heavy battle tanks, thirteen Sturmgeschütz assault guns, one Il 2, a Yak 5 and over 80 vehicles of the advancing Red Army out of action. On a single day, twelve pilots of SG 9 were able to knock out 46 T 34 tanks with their Hs 129's. The most successful was Hauptmann Ruffer, who took the number one spot in April 1943 by destroying 63 tanks. The 10.(Pz)/SG 9 marked up 100 T 34's in March alone, 30 Sturmgeschütz assault guns, a few hundred vehicles as well as six aircraft. Two aircraft were shot down by the enemy and impacted the ground burning.

Based on the successes achieved against tanks with the Mk 101/103 armament, Henschel was contracted to re-arm 12 prototype aircraft (B-3) for the time being with the BK 7.5 cannon. The B-2 and C-1 type aircraft used up to this point were formerly only equipped with the 50mm gun. Factory tests lasted for months as a result of numerous modifications. It was not until summer when the testing in Rechlin and Tarnewitz began. From there, the first three prototype aircraft went to the Erprobungskommando (EKdo) 26, where the problems concerning ejecting of spent casings continued. After 24 firings on a captured tank, RLM ordered the development stopped in October of 1944, because the 75mm weapon was not fully developed enough for the Hs 129. The Hs 129 B-3 did not was not deployed on the Front again.

Up to the beginning of 1944, 754 Hs 129's had been delivered, of which 244 were still on hand on March 30, 1944. A total of 1160 Hs 129's were produced up to the fall of 1944. The last of these aircraft (B-2/C-1) were still being used in Luftflotte 1's zone in March of 1945.

Above: Front view of the Hs 129. The external sight is easily visible as is the muzzle of the MG 151/20. The photo also gives an appreciation for the cramped conditions of the cockpit.

Below: Prototype installation of the BK 3.7 on an Hs 129 B-2 during weapons testing in Tarnewitz. It was then operational with IV./SG 9. This weapon was used to a great degree of success on the Ju 87 G.

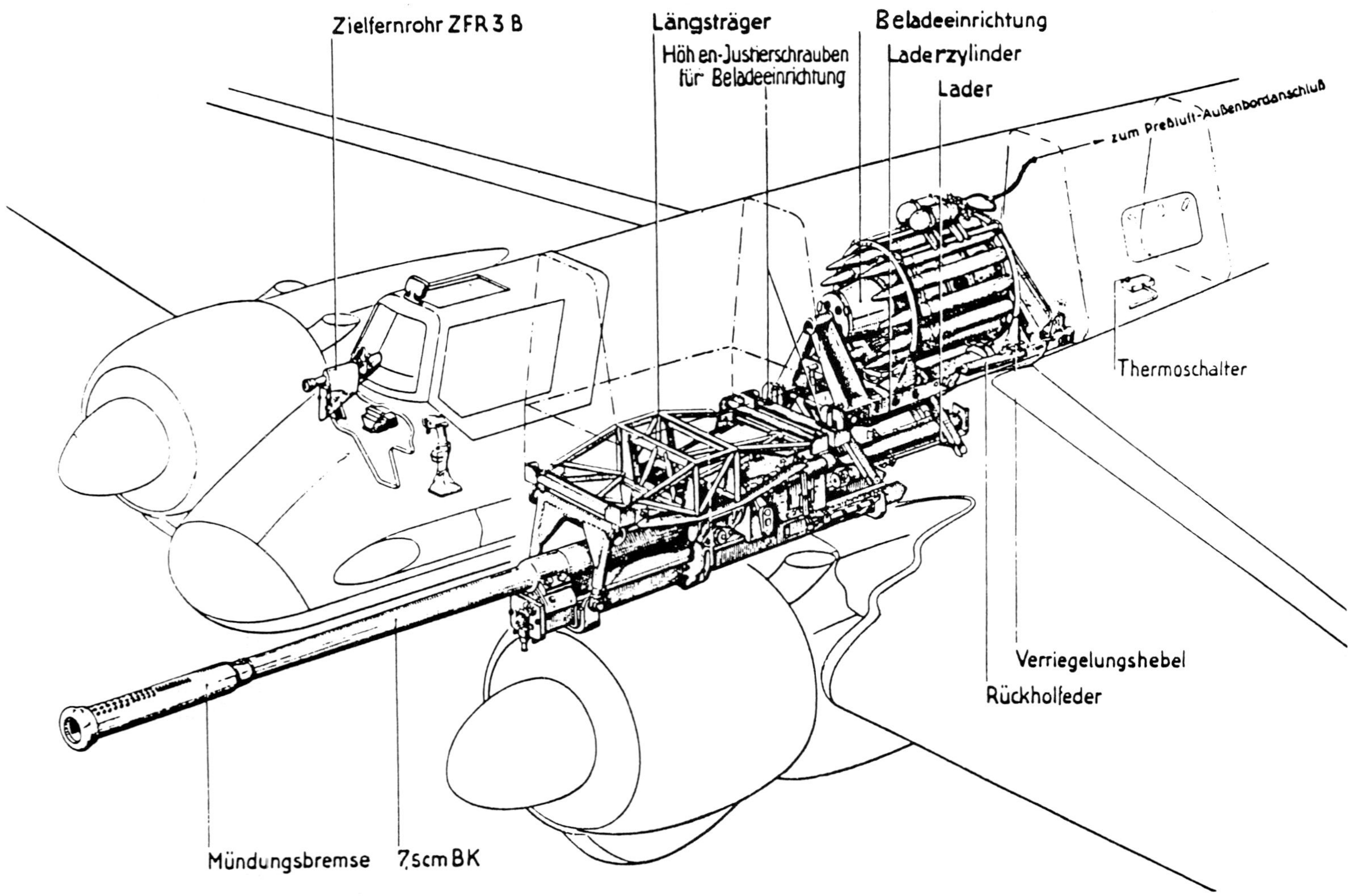

7,5 cm BK mit Beladeeinrichtung an Hs 129 B-3
(Waffenverkleidung abgenommen)

Below: In 1944, Henschel was contracted to install the BK 7.5 in twelve Hs 129 B-2's. The corresponding aircraft nomenclature was Hs 129 B-3. Due to various technical problems, this weapon was not utilized on the Front.

He 177

The RLM had planned to convert a series of twelve He 177 A-1's into "Zerstörer" (destroyer) aircraft in the early summer of 1942, each having two MK 101 cannon. The aim of this project was the engagement of enemy submarines and American four-engine bombers on their way to England. The prototype aircraft were to be tested beginning in June 1942 at Erprobungsstaffel 177. Because there was no longer any available room there, the prototype aircraft went to I./Fernkampfgeschwader (FKG) 50 in Rechlin-Lärz in the middle of October. From that point on, the aircraft remained for long periods of time on airfields in central Germany, or more accurately stood there for long periods of time, as their use over the Atlantic would not take place and those aircraft which remained were to be utilized for engaging enemy rail traffic in the East. In view of the size of the He 177 A-1, the Luftwaffe leadership soon came to the conclusion to convert these "destroyer aircraft" to normal level-flight bombers. The aircraft were then reassigned to FKG 50 and used up there. Additional tests with 30 vertically-firing 21mm cannon were not intended for engaging ground targets, but rather allied bombers within the framework of defending the Reich's airspace.

Below
In the early summer of 1942, RLM had twelve He 177 A-1's rearmed with two Mk 101's each, creating the so-called "Zerstörer" model. The result was a loss of 30km/h airspeed. Engagement of American four-engine bombers over the ocean, which was this aircraft's mission in life, was not possible because long-range fighters protected the bombers. As a result, the aircraft were to be utilized on the Eastern Front. The aircraft pictured here was the first to be converted.

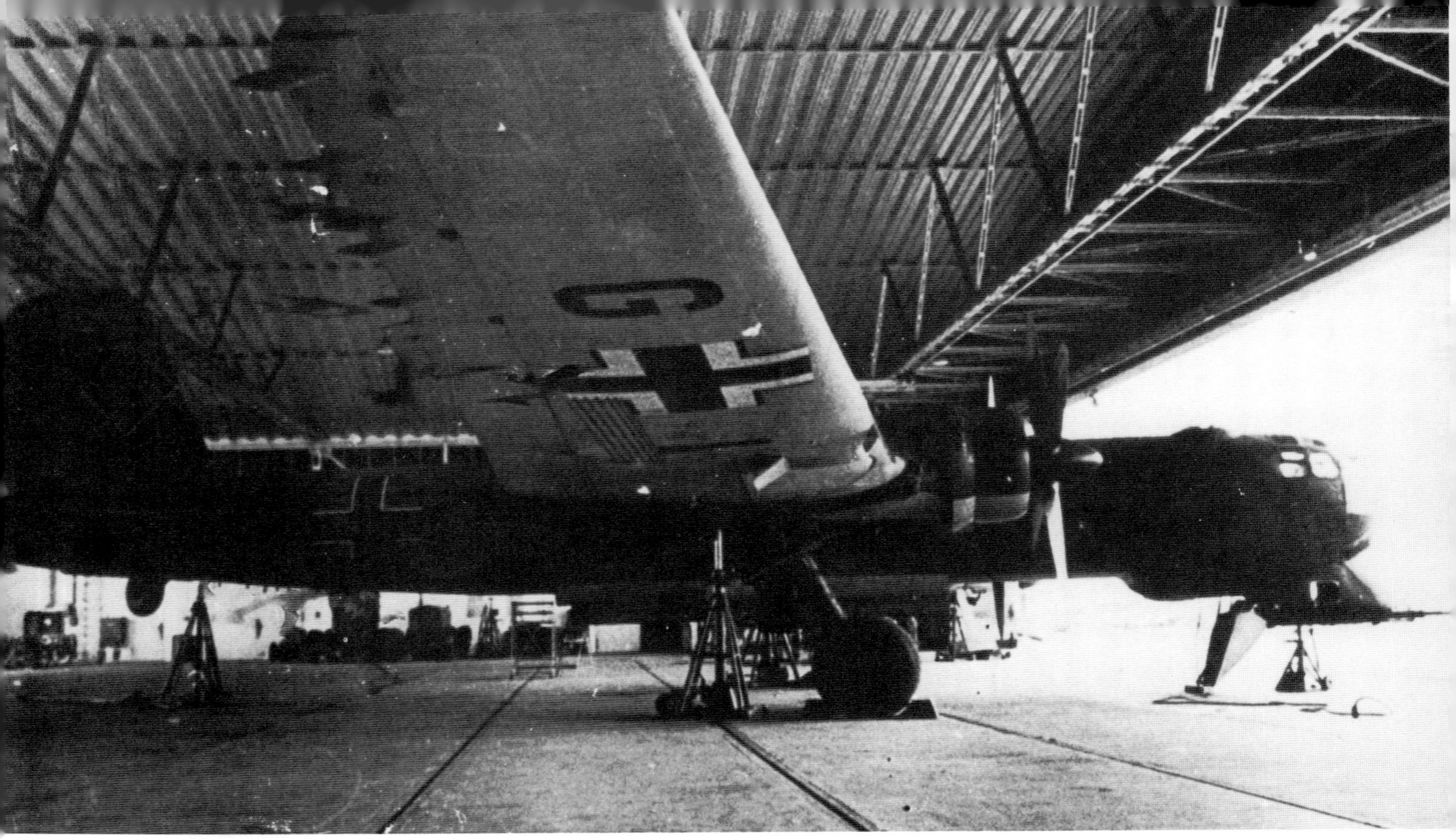

Below: This photo clearly shows the installation consisting of two MK 101 cannon, contrary to widespread literary opinion.

Above: The Fiesler Fi 156 "Storch" was one of the most adaptable aircraft of the Luftwaffe, and it was even used as a makeshift fighter. This version was designated as the "Polizei Storch" [police Storch] and fought in the struggle against partisans, among other missions.

Right: Ground crew members slinging a fragmentation bomb beneath the wing of the Fi 156.

Makeshift Aircraft

Ar 66, He 45/46, Go 145 and Si 204

Originally intended for a low-altitude reconnaissance role, both Heinkel aircraft were also assigned close air support missions even before the beginning of the war.

In particular, the remaining He 46 D-1's were repeatedly assigned to newly formed Nachtschlachtgruppen from flight schools, where they achieved astounding feats under poor weather conditions. A similar fate befell the Ar 66 and Go 145 training aircraft. The two-seaters received a temporary bomb rack for small bomb loads. Most of the two-seaters even flew of 1944/45 on the Eastern Front. Of the Ar 66, there were only 29 aircraft remaining on January 30, 1945, and 72 "Nachtschlächter" versions of the Go 145.

Two Go 145's were damaged at 2./NSG 5, one on February 1, 1945, the second on the 19th, as a result of engine failure and a forced landing. Another Go 145 was lost to anti-aircraft fire on 14 February. A fourth was lost when a bomb loosed itself from its rack and exploded while the aircraft was attempting to land. On May 3, 1945, portions of NSG 5 were located in Oberösterreich, NSG 7 was disbanded in Agram and Graz while NSG 4 had withdrawn to Olmütz-Süd. Just prior, NSG 4 had utilized several Si 204's in the area of Ratibor. A few of these aircraft also belonged to NAG 3 and fulfilled their armed reconnaissance duties quite well. On April 9, 1945, in the course of their reconnaissance of the Breslau area, three Se 204's, several Fw 190's and four Fi 156's not only performed the mission but also engaged the Soviet units with guns and a few bombs. The Fi 156, a multi-purpose reconnaissance aircraft, had been fitted with droppable munitions on an experimental basis. It had a short-term test period with troop units, when it was called the "Polizei-Storch" but was halted because the aircraft was deemed too susceptible to fire. In addition to the above aircraft, the Ar 79, Fw 44 and various Bücker training aircraft saw service as makeshift combat aircraft.

Fiat Cr 42

In addition to a good 100 Ar 66 and Go 145's, another 80 night fighters flew their missions during the last five months of the war. At the beginning of 1945 there were a total of more than 400 night-fighter aircraft, of which 225 were Ju 87 D-1's to D-5's. A few Fiat CR•42's were also in this category and served with NSG 7, which flew countless low-level attacks in February and March 1945, during which at least ten of these aircraft were lost.

Right: The Arado 66c was planned as a trainer and was altered to a makeshift bomber by temporarily hanging bombs from it beginning in 1943.

Left: The Go 145 was also provided with droppable munitions systems for employment as a ground attack aircraft.

Right: These two Go 145's are painted in the camouflage typical of nightfighters. These aircraft were operational on the very front lines right up to the war's end, which is testimony to their robust construction.

Left: American troops inspect a captured Go 145.

Anti-tank Aircraft with Rockets

Fw 190 with the Panzerschreck

In August 1944, the first Fw 190 F-8 firings system for the Raketen-Panzergranate 4322 ("Panzerschreck") anti-tank rocket arrived at the EK 26 testing facility. These were 8.8cm firing tubes for rocket-propelled anti-tank grenades.

Good results were achieved during the first 40 test firings, when six rockets were fired on each pass. Later, individual firings were conducted to determine the lead angle. Beginning at the end of September, the aircraft were sent to 5./JG 77 where training against tanks and low-altitude attacks was conducted at Udetfeld in Oberschlesien. On October 1, 1944 the squadron was deployed to its combat area. A little while later, technical testing was concluded at Udetfeld.

Starting in October, 22 Fw 190's were armed and the armorers were assigned.

Between December 11-17, 1944, night-fire tests were conducted by EK 26, but the tracer effect was not as successful as with the "Panzerblitz." At the end of January 1945, 6./SG 3 and 5./SG 77 were both deployed with the Panzerschreck. On March 22 1945, Fw 190's of 8./SG 1, 6./SG 3 and 5./SG 77 flew with the Panzerschreck. Other firing systems could be found with 2.(SG)/9 and 10.(Pz)/SG 77, which had Fw 190's and Ju 87's in their inventories. As opposed to the "Panzerblitz," no other units were converted to the "Panzerschreck."

During the last low-level attacks, which also saw AB 250 and 500's and some fighter squadron pilots with their Me 262's taking part, some of the Fw 190 F-8's were deployed at the end of April with Panzerblitz weapons to relieve the Reich's capital city.

Fw 190 with the Panzerblitz

In addition to the rocket tubes known as the Panzerschreck, the Panzerblitz was also developed. In mid-September 1944, the first Fw 190 F-8 was fitted with Panzerblitz lattices and testing of the sighting and lead angle was begun as quickly as possible. Twelve test-firings were conducted with the "Panzerblitz" up to September 19, 1944. At the end of that month, captured enemy tanks were fired upon at an angle of 30 degrees. Aside from the tanks, a dummy ship and a locomotive were also used as targets. During testing conducted by III./SG4, good results were achieved during 42 attacks. Practically foolproof versions of the "Panzerblitz" system, which had been manufactured by the Heber firm, arrived in Udetfeld on November 11 1944 and were mounted right away. At EK 26 in early December 1944, the training of pilots from III./SG 4 and a squadron captain from SG 151 was the order of the day.

Personnel of II./SG 1, II./SG 2, II. and III./SG 3, II./SG4 I./SG 10 as well as SG 151 were intensively trained up to December 11 1944 and 82 aircraft were converted to carry the improved "Panzerblitz 2" (this was the modified R4M rocket with the shaped charge warhead).

A few days later, night fighting with the new device was tested, during which the hits on target could be easily observed.

Further conversions and training were lasted a few more weeks so that at the beginning of January 1945, the first live deployments could take place. The rocket armament was used for the first time by III./SG 4 in the Luftflotte 6 sector. In the time between January 21 to March 16, 1945 the weapons were deployed on a total of 16 days. In 25 sorties conducted during that time, a total of 115 Fw 190 fired exactly 934 "Panzerblitz" rockets. The pilots reported 23 enemy tanks as destroyed, another eleven were rendered immobile. The Panzerjäger aircraft were given protection by fighter planes deployed in approximately the same strengths.

Soviet units reacted immediately to this new threat. The lead elements of the armor columns now were accompanied by anti-aircraft and self-propelled artillery. In silence, the Soviet units would try to hide in forests and among houses or dug their tanks in.

In March 1945, the conversion of the Fw 190 F-8 was the center of effort in Gardelegen. Pilots and weapons technicians were also trained there, for which enough fuel was specially set aside.

At the end of the month there were "Panzerblitz" equipped Fw 190's at 1. and 3.(Pz)/SG 9, 9./SG 77 and at three additional squadrons within Luftflotte 6's sector.

The number of combat-ready units climbed to 12 by April 20, 1945. Due to the situation of the war, there were very difficult transportation problems, such that tens of thousands of projectiles which were produced could not be transported to the line units.

Due to the high rate of accuracy of the "Panzerblitz" weapon, the commander of the test facilities was prompted to order the SG 113 project started in the first days of April to be abandoned. Nevertheless, he did conduct tests with various explosive warheads and sent his free personnel to combat units.

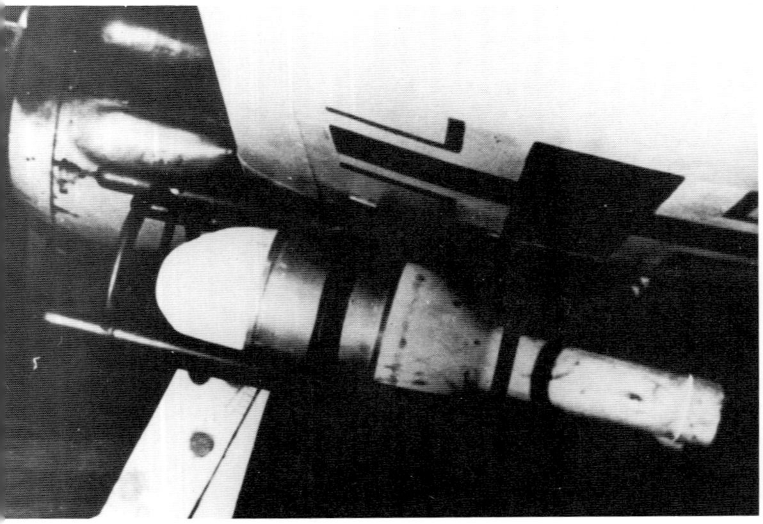

Above: Testing was conducted with the WGR 28cm grenade launcher on the Fw 190. Due to the low speed required and the weapon's lack of accuracy, this weapon was deemed as not airworthy.

Above: Installation of the SG 113 A "Förstersonde" on an Fw 190 F-8. Due to problems with the firing priority, this weapon was not utilized on the front. The SG 113 was also mounted on three Hs 129 B's and, on 18 January 1945, two successful flight tests were made.

Above: This Fw 190 F-8 is equipped with rocket firing tubes for engaging enemy tanks.

Above: Parked ETC-equipped Fw 190 G's to be deployed as fighter-bombers on the Eastern Front.

Below: These Fw F-8/R-1's of I./SG 51 scuttled on 22 January 1945 at Oels airfield. The aircraft were fitted with firing tubes for the Panzerschreck.

Below: Eleven single-shot firing devices of the RZ 65 air-launched rocket beneath the fuselage of a Bf 110. The projectiles were intended for engaging air and ground targets alike.

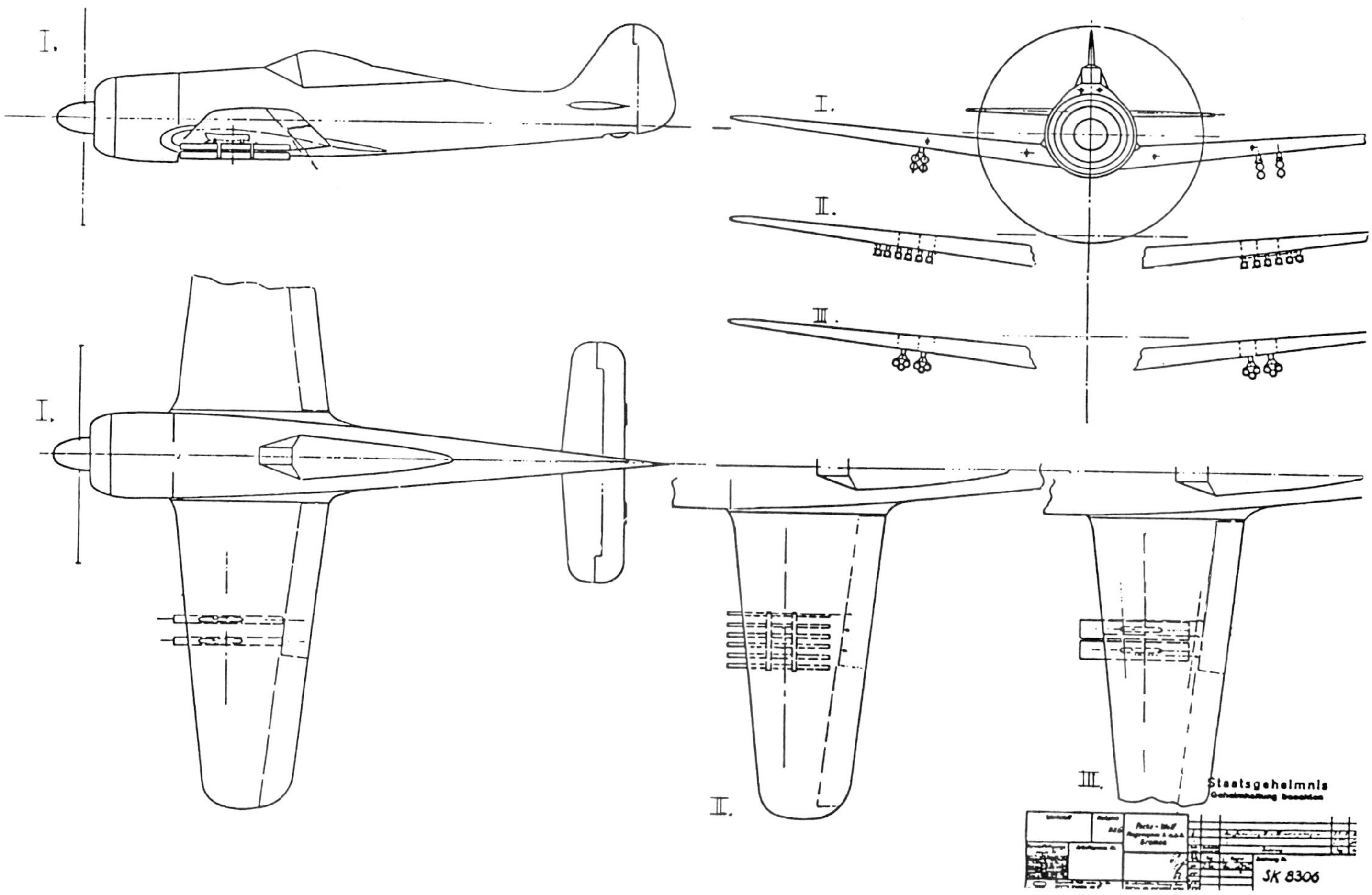

Above: Location of the individual weapons on the Fw 190.

Diagram key:
I. installation of the PD 8.8 (individually and in bundles)
II. installation of the Panzerblitz
III. installation of the Panzerschreck

Below: Operations schematic for the SG 113.

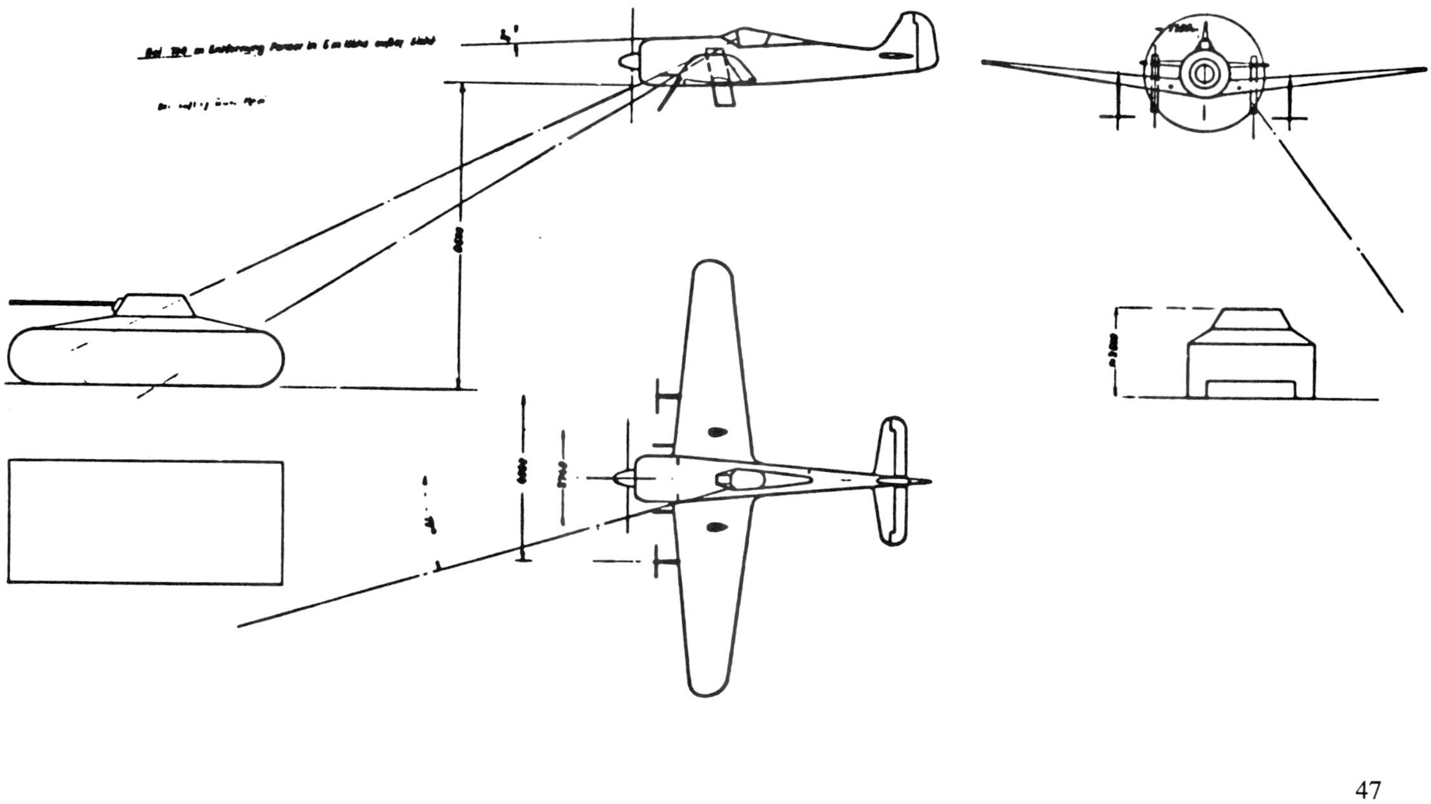

Junkers EF 126

The "Panzerblitz" was also intended weaponry for light combat aircraft such as the Ju EF 126 "Elli," and not just for the Fw 190. This was a single-seat aircraft propelled by an Argus rocket engine with a performance of 500kg thrust and launched from a rail or a skid chassis. The aircraft, which was 8.0 meters in length and weighed about 2,800 kilograms, was to be armed with two MG 151/20 machine guns with a 180-round supply of ammunition apiece. Additionally there were plans for two AB 250's or four AB 70's and twelve "Panzerblitz" rails. The development of the "Elli" was continued in Dessau after the war under Soviet direction. Five prototypes were produced before April of 1946. After test-firings had been conducted with the 20mm weapon in December 1945, plans for installing rocket systems in the wings were made. What ever became of the project remains unknown.

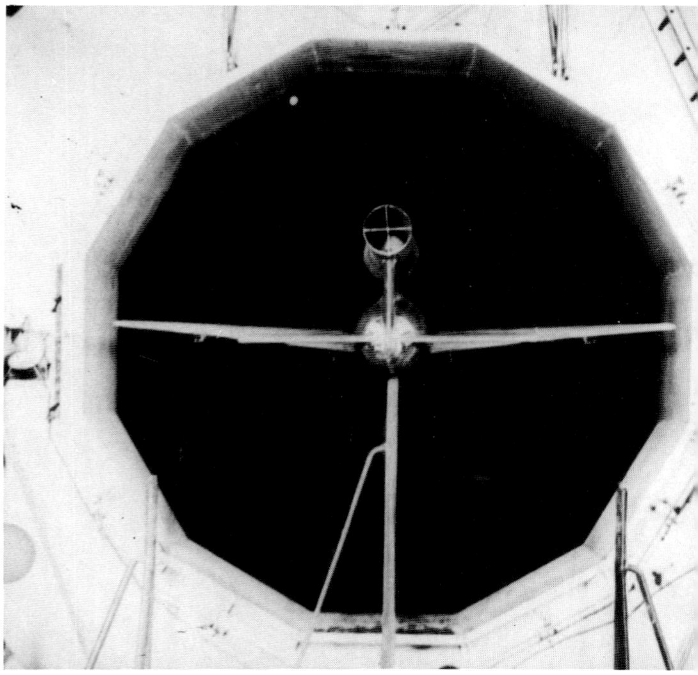

Above and left: This version of Junkers project EF 126 was designated as a ground attack aircraft with the Argus jet engine. The design underwent wind tunnel testing in January of 1945.

Below: Based on its performance, RLM ordered the conversion of the Me 262 A-1a to the A-1a/Bo "Blitzbomber" in keeping with the Reichsmarschall's address on the Whitsun holiday. Maintenance of the jet fighters was assumed by Erprobungskommando Lärz beginning in June 1944. The landing gear posed problems as it was not designed to carry the heavy weight of the bomb loads.

Single-seat Jet Models

Blitzbomber Me 262 A-1/2

The "Blitzbomber," later called the Me 262 A-2, received a greater priority after the speech given by the Reichsmarschall on the occasion of the Whitsun holiday in 1944. Its application as a fast attack airplane lead to the inclusion of the Blitzbomber, or lightning bomber, in the Me 262 emergency research program which involved E2 through E8 units in Reichlin. One difficult point was, quite naturally, uncovered during the bomb testing (done by E7) in Lärz near Reichlin. The test facility in Lärz (EKL) assumed the maintenance responsibilities of the Blitzbomber which arrived there on June 10, 1944. A further twelve such aircraft landed there up to early July 1944. Due to difficulties with the landing gear, testing was delayed about four weeks. Thanks to quickly enacted improvements to the landing gear, the takeoff weight of the aircraft was increased to 7,100kg. Finally, between early August and September 20, 1944, about 800 sorties were flown with various loads. During this time, aircraft readiness reached a maximum of about 70%. Thanks to the improvement made to the throttle damper, the hydraulic systems as well as both of the Jumo 004 turbine engines, a deployable version was soon available which could also be successfully flown in unfavorable weather conditions.

The EKL conducted tests on the droppable munitions both on the ground and in the air. During this time, the Messerschmitt bomb rack, which was basically was nothing more than a fared and easily altered ETC 500, underwent an evaluation and had improvements made. Afterwards it was possible for the aircraft to carry not only the SC/SD 250, SC/SD 500, but also the Abwurfbehälter 500-1 and 500-2 droppable containers. The BT 200 torpedo could also be carried, but the BT 200 did not quite fit as it stretched out over the landing gear doors. While problem details were still being worked on in Rechlin-Lärz, Kommando Schenk (the operational element of KG 51) conducted the first live deployments of the Blitzbomber. Here, eight Me 262's attacks targets on August 25, 1944 along the Seine river near Nantes. Each of the aircraft carried AB 500 with SD 10 charges. There were, however, hardly any attacks flown against armor columns as the priority lay with attacking enemy supply lines.

After seven missions, the unit had to transfer from Chateaudum to Belgium. During its forced retreat to Rheine/Westfallen in September 1944, the unit had eleven Me 262 bombers. The first mass strike conducted with these aircraft came on September 26, 1944 when 20 Me 262's of 3./KG 51 dropped AB 500 bombs on the area of Nymwegen. Pilots of Schenk EK also conducted reconnaissance flights with their twelve Me 262's. Despite some losses in October 1944, enemy airfields were attacked during poor weather. KG 51 also flew many harassing missions against targets in Belgium and the Netherlands. More than 30 Me 262 A-2's were mustered for low-level and ground attacks beginning in November of that year. One of the most massive attacks was conducted on February 22, 1945 when 48 Blitzbomber's flew under the cover of 87 piston fighters in the West. The last attacks on the heavily contested Bridge at Remagen on March 15, 1945 and on the advancing allied armor and motorized units in the area of Bad Kreuznach permitted little relief for the pilots. The I. and II. Gruppe lost more and more pilots because of this and the far superior American P-47 and P-51 fighters.

In the beginning of May 1945, pilots of JG 7, HV 44 and of KG 51 flew their last missions from Prague-Ruszin.

Below: The first attack testing with the Me 262 A-1a/Bo was conducted in Lärz in August 1944. This photo shows ground crew loading the Me 262 with SC 250 bombs.

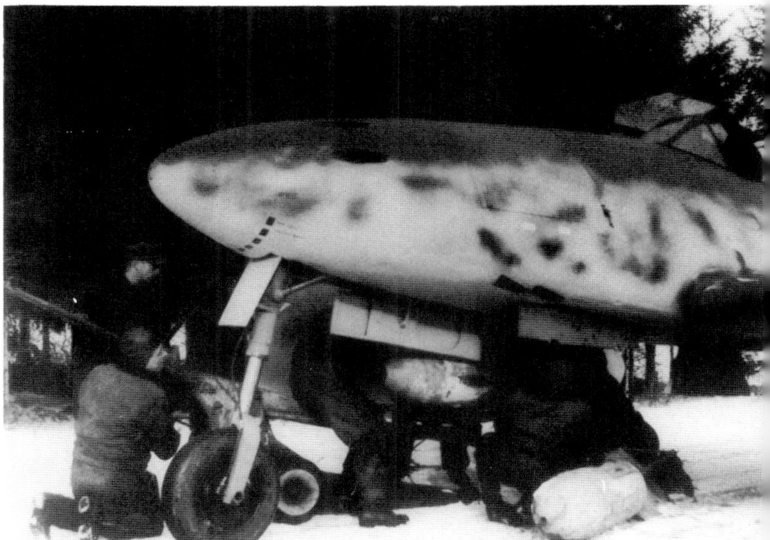

Me 328

In December 1942, the Technisches Amt awarded a contract for a total of 30 Me 328's to be built in the most simple form. RLM order that in addition to the ten test aircraft of the A-1 series (Me 328 V1 to V10), which were to be produced at DFS, another 20 aircraft were to be manufactured. These aircraft belonged to the planned Me 328 B-1 series (V11 to V30) and were to be manufactured by the Jacobs & Schwayer. After comprehensive research and wind tunnel testing in 1943/44, the fast attack bomber Me 328 B-1 received even more emphasis because it could be built in a simplified version of the A-1 model made of wood. With the exception of a few test prototypes, which were partly built with Argus steel tubing, a large part of the expense was for nothing as the Me 262 bombers were given more and more importance.

Above: The Me 328 as a testbed with two 300kp AS 014's.

Below: RLM ordered the construction of ten Me 328 A test aircraft by DFS. These were tested in tow, mounted on a Do 217 E, but without engines.

Below: Because of the drive for an easily-produced airplane, the Me 328 B came into being as a light high-speed fighter. This aircraft was to have a range of 555 kilometers with a 1000kg bombload. This according to the pre-construction description from Messerschmitt dated 15 December 1942. Its speed at 1000 meters altitude was 790km/h without external weapons stores.

Below this is a wind tunnel model of the Me 328 B of the DFS Institut A in Darmstadt. It had a longer fuselage, slightly altered wing roots and fuel tanks on the wings.

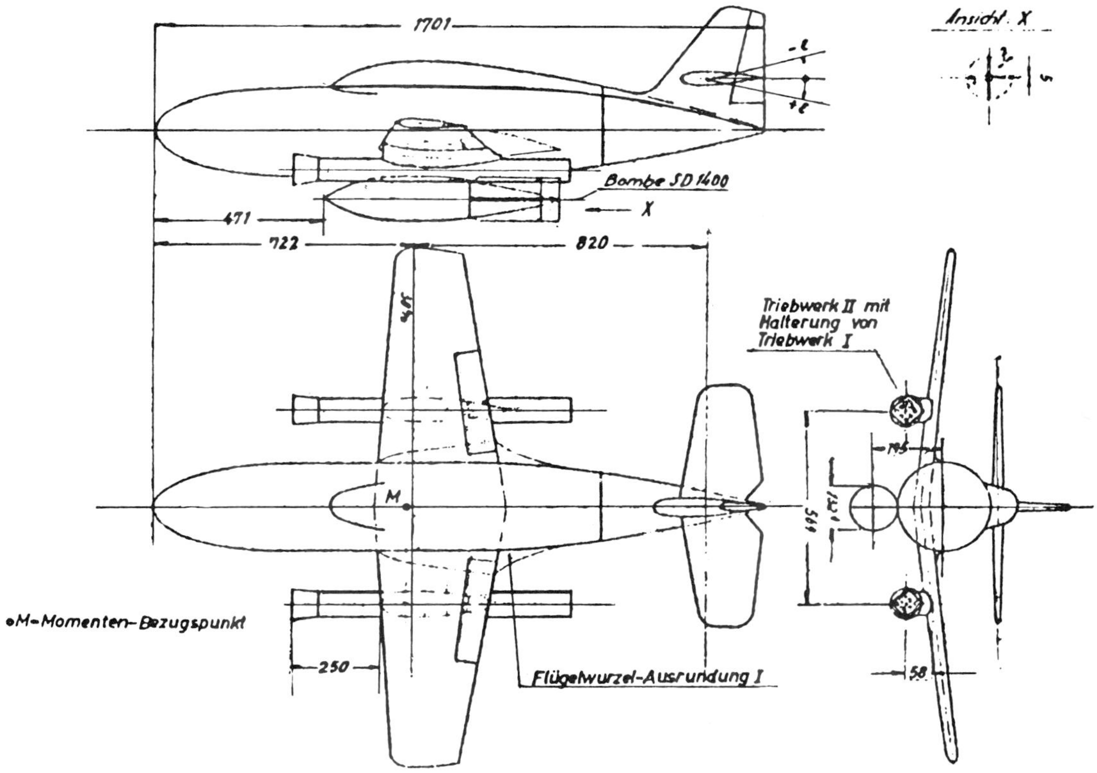

Above: A model study for the Me 328 B with wing-mounted Argus jet engines and SD(Top) 1400 bombs.

Above: Photo of the Me 328.

Hs 132

With the publishing of RLM (the GL/C-E2 department) order dated January 31, 1944, Henschel was officially directed to rename its projekt "P 123" to Hs 132.

However, because of a lack of production capacity, the project did not reach manufacturing stages until May. On July 25, 1944, fifty percent of the work had been completed and detail work was started. After some delays, the Hs 132 V1 fuselage assembly was begun in September. After some alterations to the external load hard point mounts, and as a result of reinforcement work done on the frame, the delivery date was pushed back to January 1945. In October 1944, the first BMW 003 engine arrived at Henschel. Four weeks later, the fuselage was ready for final assembly, but completion of the tail assembly was delayed from month to month.

Production of the Ta 152 wings and other subcontract license work as well as the continued development of rockets impaired the assembly of the Hs 132, which, as a fighter-bomber, was to be equipped with the most modern air to ground weapons. The aircraft was to be used in dive-bomb attacks, with inclinations up to 70 degrees and could reach speeds up to 910 kilometers per hour. AVA Göttingen tested the a 1:6 model of the future Hs 132 fuselage in a wind tunnel. Up to mid-March, only the main aircraft portions of the first prototype were assembled. The aircraft never saw final assembly. The only published photograph of the Hs 132 is simply a photograph montage. The aircraft's parts as well as all available files were plundered by the Soviets.

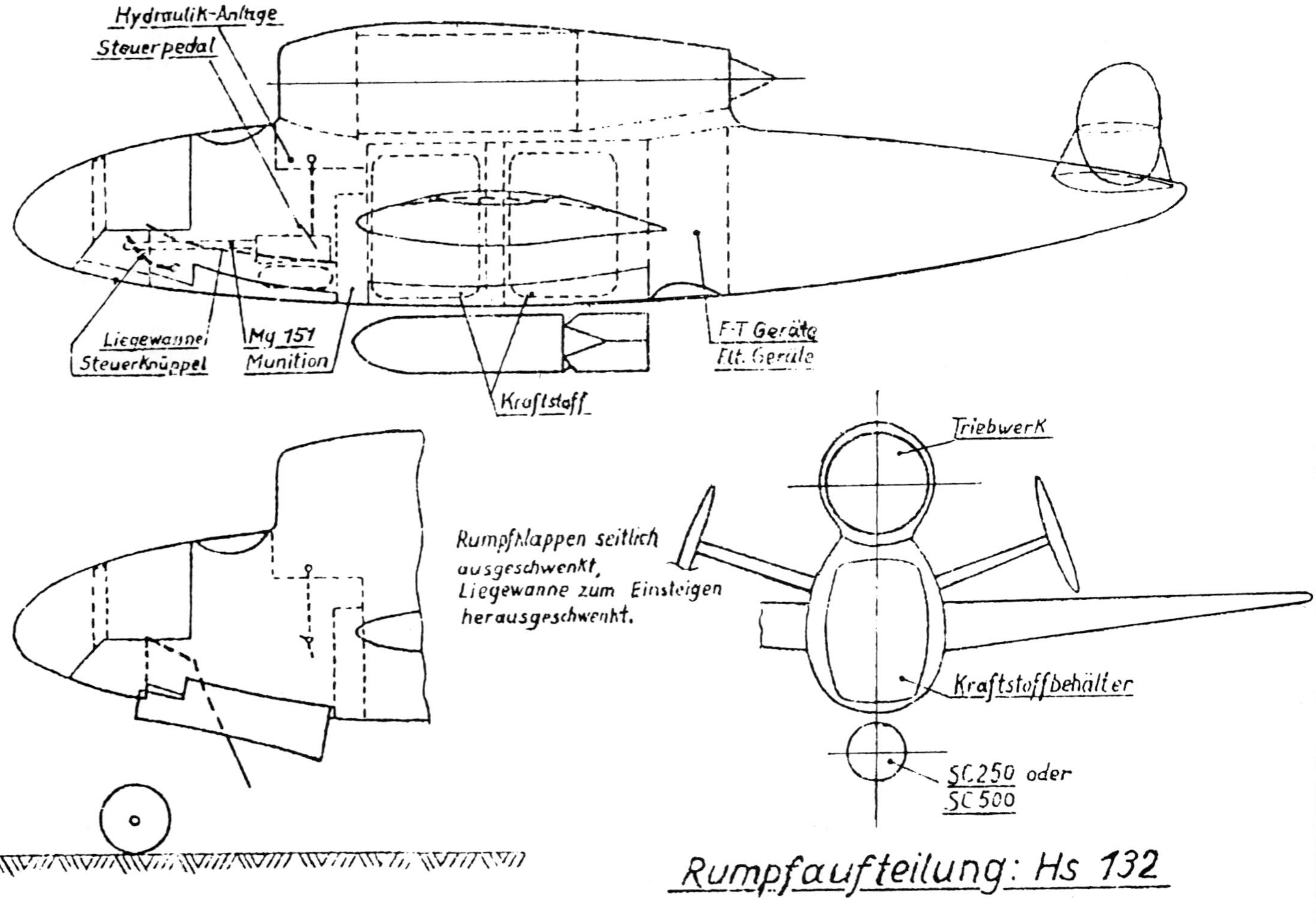